Springer Series in Advanced Manufacturing

Other titles in this series

Effective Resource Management in Manufacturing Systems: Optimization Algorithms in Production Planning
Caramia and Dell'Olmo

Advances in Design
H.A. ElMaraghy and W.H. ElMaraghy (Eds.)

Condition Monitoring and Control for Intelligent Manufacturing
Wang and Gao (Eds.)
Publication due: January 2006

Brahim Rekiek and Alain Delchambre

Assembly Line Design

The Balancing of Mixed-Model Hybrid Assembly Lines with Genetic Algorithms

With 95 Figures

Brahim Rekiek
Alain Delchambre
Université Libre de Bruxelles
CAD/CAM Department
50 av. F. D. Roosevelt
CP 165/14 1050 Brussels
Belgium

Series Editor:
Professor D. T. Pham
Intelligent Systems Laboratory
WDA Centre of Enterprise in
Manufacturing Engineering
University of Wales Cardiff
PO Box 688
Newport Road
Cardiff
CF2 3ET
UK

British Library Cataloguing in Publication Data
Rekiek, Brahim
Assembly line design : the balancing of mixed-model hybrid
assembly lines with genetic algorithms. - (Springer series
in advanced manufacturing)
1. Assembly-line methods 2. Genetic algorithms
I. Title II. Delchambre, A.
670.4'2

Springer Series in Advanced Manufacturing ISSN 1860-5168
ISBN-13: 978-1-84996-555-2 e-ISBN 978-1-84628-114-3 Printed on acid-free paper

Printed in Germany

9 8 7 6 5 4 3 2 1

Springer Science+Business Media
springeronline.com

This book is dedicated to my parents,
to my wife Aaida, and
to my children Saad and Inas.

Dr B. Rekiek

Foreword

This new book 'Assembly Line Design' by Dr Brahim Rekiek and Professor Alain Delchambre is an important contribution in this domain. Its interest is in an integrated approach to the preliminary design of assembly lines (ALs). This approach is based on the grouping genetic algorithm (GGA), where the logical layout (LL) is designed to consider all the constraints and specificities of real-life manual and hybrid multi-product ALs. The LL is defined as the balancing and the resource planning. In addition, a new approach based on multi-objective GGAs is developed which includes the branch-and-cut algorithm combined with a multi-criteria decision-aid method. In this book, the logical and physical layouts are treated simultaneously. First, tasks (that perform activities) are grouped together in workcentres. Second, tasks are assigned to stations. The new concept of 'balance for operation' is introduced to deal with the changes during the operation phase of ALs. This concept permits one to treat balancing and scheduling at the design stage.

The authors have a great experience in practical AL design and balancing. Their scientific publications are well known and widely cited. Undeniably, this new book offer new vision and perspectives for development of industrial research and engineering methods for AL design. It provides a systematic analysis, efficient engineering concepts, and techniques to handle this design problem. It is a pleasure to foreword this excellent book as an important source for researcher, industrial engineer, faculty staff and graduate students in industrial engineering, management science, operations research and mechanical engineering.

Professor Alexandre Dolgui
Ecole des Mines de Saint Etienne, France
May, 2005

Preface

The design process is traditionally a time-consuming and an iterative business. First, a preliminary design is created, analysed, and then experimented to determine its quality. The process of search and evaluation is repeated until the design is viewed as being acceptable. Computer-aided design (CAD) software, simulation and analysis tools are widely used today. In contrast, automatic design techniques are less common. The recent success in design is due to the adaptive search techniques, in particular the genetic algorithms (GAs). GAs are powerful and broadly applicable stochastic search and optimisation techniques. They are the most widely known kind of evolutionary computation methods.

Assembly lines (ALs) are production systems composed of a succession of stations, connected by a conveyor, performing a set of tasks on the product passing through them. A production workshop can be set up following various topologies (*e.g.* lines, cells, combination of several lines, *etc.*) The line layout problem is composed of a logical and physical layout. The logical layout is defined as the AL balancing (ALB) and the resource planning (RP) problems. The ALB is used for manual ALs and it aims to balance loads of stations. For hybrid ALs (manual, robotic and automatic tasks), RP assigns resources to tasks and assigns tasks to stations. The physical layout determines the space requirements taking into account station dimensions, material storage, *etc.* The aim is to minimise the total cost of the line by integrating design (space, cost, *etc.*) and operation issues (cycle time, precedence, availability, *etc.*).

AL design (ALD) problems often have a complex structure due to multiple components (*e.g.* tooling, material handling facility, line efficiency, cost, imbalance, reliability, stations space, *etc.*). A number of design alternatives may exist. The problem can easily become unmanageable if the designer has to consider all these alternatives. Thus, many practical search and optimisation problems are considered as multiple objective problems (MOPs) and require a compromise among conflicting objectives. Since it is impossible to replace designers experience and creativity, it is important to support them with a set of tools to investigate and propose solutions. Using this information, the designer tests some alternatives and makes his decisions. Owing to the difficulty of ALD problems, metaheuristics are often used.

In applying GAs to solve MOPs one has to deal with the twin issues of searching a large and complex solution space and at the same time dealing with multiple and conflicting objectives. Selection of a solution from a set of possible solutions on the basis of several criteria is considered as a difficult task. Some methods reduce the problem to a mono-criterion one (weighted-sum approach). Other studies adopted the Pareto-based GA technique. The main drawback of Pareto approaches is the number of solutions the decision maker (DM) has to choose among them. The user cannot easily decide among more than a few solutions.

We present a new multiple objective grouping GA (MO-GGA) which is based on the GGA and multi-criteria decision-aid (MCDA) method called PROMETHEE II. The GA iteratively samples the trade-off surface (Pareto) while the MCDA method narrows the search. The choice of a solution over the others requires knowledge of the problem. It is the task of the DM to adjust the weights for guiding the algorithm to find good solutions. Optimising a set of objectives has the advantage of producing a single solution, without any further interaction by the DM.

In order to deal with line balancing, a new algorithm called 'equal piles' for ALs based on the so-called 'boundary stones' is introduced. The hard constraint is the fixed number of stations (piles) and the aim is to find the best balanced assembly system. In the case of the RP, the aim is to select equipment to carry out the assembly tasks. We present a new method which is based on the MO-GGA, the branch-and-cut algorithm followed by the MCDA method. To deal with the changes during operation phase of ALs, a new concept of balance for operations is introduced. The balancing of ALs is mostly uncoupled from the facility layout problem which yields sub-optimal line layouts. An iterative procedure is proposed to treat the two problems partially at the same time. First, tasks that perform similar activities are grouped together in a workcentre. Then, for each workcentre, tasks are assigned to stations. The main concern of this approach is the quality of the resulting line in terms of balancing and its suitability to the material flow requirements of the production system.

The last part of this book is dedicated to an integrated method of designing ALs. The software OptiLine is developed at the CAD/CAM Department of the Université Libre de Bruxelles, Belgium.

Acknowledgements

Several people contributed to this work. Many thanks to Dr E. Falkenauer, Dr F. Pellichero, Dr P. De Lit, A. Rekiek, Dr H.A. Saleh and Dr O. Bouhali.

Special thanks to Professors A. Dolgui (Ecole des Mines de Saint Etienne, France), P. Gaspart (Université Libre de Bruxelles, Belgium), J.-M. Henrioud (Laboratoire d'automatique de Besanon, France), B. Raucent (Université Catholique de Louvain, Belgium), and B. Mareschal (Université Libre de Bruxelles, Belgium) for their fruitful comments.

Our thanks to Dr E. Falkenauer, General Manager of the Optimal Design company. He provided us a real-world case study illustrating the concepts described in this book. The case study has been optimised using the OptiLine software package which has been developed by Optimal Design.

The financial support of the Région Wallonne through a project entitled CISAL is acknowledged. My thanks to the Université Libre de Bruxelles, Belgium, especially to the staff of Service de Mécanique analytique et CFAO.

To our families with ultimate respect and gratitude for their continuous support. Many thanks also to the many interested readers of our research papers for some stimulating discussions at conferences, workshops and over the Internet.

Dr B. Rekiek
Professor A. Delchambre

Contents

Part III Assembly Line Layout

List of Abbreviations

AI	Artificial intelligence
AL	Assembly line
ALB	Assembly line balancing
ALD	Assembly line design
B&B	Branch and bound
B&C	Branch and cut
BD	Balance delay
BFO	Balance for operation
BPP	Bin packing problem
CAD	Computer aided-design
CE	Concurrent engineering
CISAL	Outils d'aide à la conception interactive des produits et de leur ligne d'assemblage
CM	Cellular manufacturing
COP	Combinatorial optimisation problem
CS	Capacity supply
DFA	Design for assembly
DM	Decision maker
DP	Dynamic programming
E	Line efficiency
EPALP	Equal piles for assembly line problem
ES	Evolutionary strategies
FABLE	Fast algorithm for balancing line effectively
FFD	First fit decreasing
FG	Functional group
GA	Genetic algorithm
GC	Goal chasing method
GGA	Grouping genetic algorithm
GT	Group technology
HAL	Hybrid assembly line
I	Line idle time

IB	Imbalance
ICA	Individual construction algorithm
JIT	Just in time
LL	Logical layout
LP	Linear programming
MAL	Manual assembly line
MCDA	Multi-criteria decision-aid
ML	Model launching
MOALBP	Multiple objective ALBP
MOB-ES	Multiple objective evolution strategy
MOEA	Multiple objective evolutionary algorithm
MOGLS	Multiple objective genetic local search
MOGA	Multiple objective genetic algorithm
MOGGA	Multiple objective grouping genetic algorithm
MOP	Multiple objective problem
MPAL	Multi product assembly line
MWkCALB	Multiple workcentres ALBP
NPGA	Niched pareto genetic algorithm
NSGA	Non-dominated sorting genetic algorithm
OGA	Ordering genetic algorithm
OMT	Operating modes and techniques
OV	Ordering variants
OX	Order crossover
PBX	Position based crossover
PG	Precedence graph
PL	Physical layout
PMX	Partially mapped crossover
PROMETHEE	Preference ranking organisation Method for Enrichment evaluations
PSGA	Problem space genetic algorithm
RD-MOGLS	Random directions multiple objective genetic local search
RP	Resource planning
RPW	Ranked positional weight
RRPW	Reversed ranked positional weight
RWS	Roulette wheel selection
SA	Simulated annealing
SALBP	Simple ALBP
SMCT	Scheduling method choice tool
SPAL	Simple assembly line balancing
SPEA	Strength pareto evolutionary algorithm
ST	Station time
SX	Smoothness index (SX)
TALB	Tree assembly line balancing
TS	Tabu search
TVR	Time variability ratio
VEGA	Schaffer's vector evaluated GA

Part I

Assembly Line Design Problems

1

Designing Assembly Lines

1.1 Introduction

Many attempts have been made in the last few years to investigate the use of *semi-automatic* methods of design as human design is time consuming. The term design implies a systematic planning processes prior to the execution of a plan in order to solve problems. Design is distinguished from other forms of planning by the level of precision used, expertise and care. It involves the consideration of many factors that may affect or be affected by the execution of a given plan. Many designers have the impression that the design is a 'cut-and-paste from old design' activities. This is not the case, as the *creativity* has a major role in design.

Assembly lines (ALs) are the most commonly used method in a mass production environment. They allow the assembly of products by workers with limited training and by dedicated machines and/or by robots. The main objective of assembly systems is to increase the efficiency of the line by maximising the ratio between throughput and cost. AL design (ALD) involves the design of products, processes and plant layout before the construction of the line itself. These different modules interact at the different stages of ALD [35]. The product analysis proposes a product design review based on the classical `design for assembly' (DFA) rules and precedence constraints between tasks. The operating modes and techniques module proposes an assembly technique and the possible modes (manual, automated, robotic) for each task. The line layout (LL) module assigns tasks to a set of stations and decides on the position of stations and the resources on the plant floor (Figure 1.1).

1.2 Assembly Line Design

The design of efficient assembly workshops is a problem of considerable industrial importance. ALs are production systems composed of a succession of

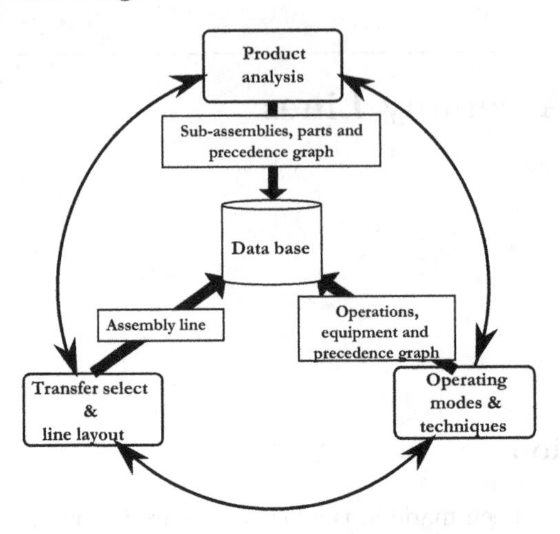

Figure 1.1. Methodology and information flow of the ALD [36]

stations performing a set of tasks on the product passing through them. The assembled product takes its shape gradually starting with one part (the base part), with the remaining parts being attached at the various stations which are visited by the product. A paced AL is a usual topology for medium and high production volumes [39]. In general, for simple products a single linear AL with possibly parallel stations can do the job. For complex products, the assembly system is mostly decomposed into sub-systems with their own cycle time, reliability, and stations requirements.

Many successful companies have adopted several working practices and tools known as concurrent engineering (CE) to improve their products' development. The main aim of CE is to integrate product and process development in order to reduce the design lead-time and to improve its quality and cost. The LL problem is known as logical and physical layout [39]. The elaboration of the *logical layout* of the line consists of distributing the tasks among stations along the line, while the *physical layout* decides on the disposition of stations, resources, conveyors, buffers, *etc.* on the shop floor. The logical LL is composed of *AL balancing* (ALB) and *resource planning* (RP) problems (Figure 1.2). The balancing used for manual ALs aims to balance the stations' workloads. For hybrid ALs (HALs) (where operations can be executed manually using robots or automated equipment) the RP assigns resources to tasks and tasks to stations. The objective is to minimise the total cost of the line by simultaneously integrating design (*e.g.* station space, cost, *etc.*), operation issues (*e.g.* cycle time, precedence constraints, availability, *etc.*) and designer desires (*e.g.* tasks complexity, *etc.*). Figure 1.3 shows the main features (blocks) of the concurrent ALD approach that will be discussed in detail in Chapter 10.

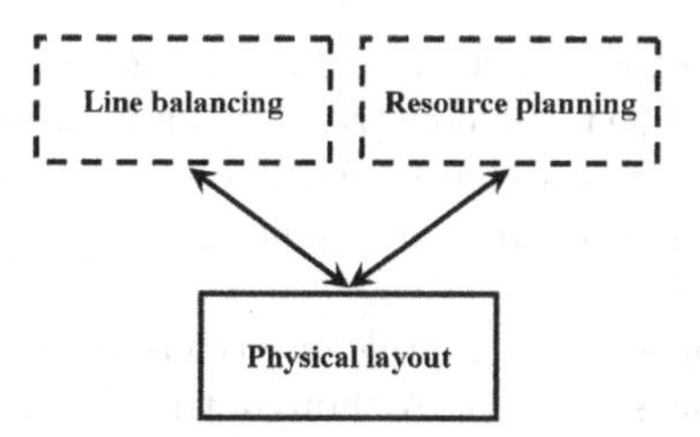

Figure 1.2. LL problem

A line design problem often has a complex structure due to multiple components (*e.g.* tooling, operators, material handling facilities, *etc.*). For a single product, a number of design alternatives may exist. The problem can easily become highly complicated if the designer has to consider all the possible combinations of these alternatives. Therefore, the problem must be solved with a structured approach. For a given product and a manufacturing environment, the design objective and constraints should be defined. A computer system which is inspired by nature (Darwinian evolution) is presented to design new ALs starting from a set of specifications. The designer system is *generic* (*i.e.* it has to be capable of evolving a wide range of different line designs with minimal reconfiguration by a designer).

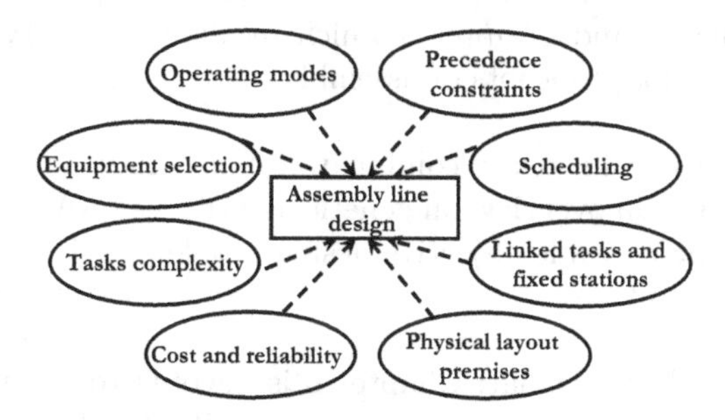

Figure 1.3. Concurrent design of an AL

1.3 Designing or Optimising?

The complexity of design is not due to the physical, material or procedural factors; rather, it depends on understanding a problem and making well-founded decisions. There are some general design steps that the designer has to follow [106]. These steps are: (1) formulating the problem to be solved, (2) breaking it down into sub-problems, (3) grouping ideas that must be discussed, (4) evaluating and redesigning (if needed) the current design, and finally (5) implementing the proposed model. In general terms, design is the process of specifying a description of an object (product, program, *etc.*) that satisfies a collection of *constraints.* The term 'constraint' usually means something which is either satisfied or not. It is a characteristic of many design problems that new constraints emerge as *decisions* are made [14, 54].

The *combinatorial optimisation problems* (COPs) are characterised by a finite number of feasible solutions. Although the optimal solution of such problems can be found by an enumeration, especially for practical problems. We can observe a tendency to use *heuristics* rather than exact methods. A *metric* is needed to identify a successful search (*i.e.* indicate if the *goal* looking for was reached or not). This metric could be *binary* ('found', 'not found yet') or be *information* on the proximity of the current solution in relation to the *best* solution. In many discrete-space problems, there is no *better* or *worse* solution, but the solution is either wrong or right. The aim is rather to find a solution that satisfies different constraints [54, 90, 152, 155].

1.4 Layout of the Book

This book is divided into four parts as follows:

Part 1 deals with ALD problems and consists of Chapter 2 which introduces design problems, and Chapter 3 which recalls the history of ALs and summarises the principal concepts of assembly.

Part 2 deals with evolutionary combinatorial optimisation and consists of Chapter 4 which gives an overview on genetic algorithms (GAs), and Chapter 5 which considers the multiple objective design problem and introduces some improvements made to GAs.

Part 3 deals with AL layout: Chapter 6 is devoted to the manual ALB problem and explains the new method 'the equal piles for ALs', and Chapter 7 is devoted to RP for HALs and shows how to tighten the gap between the academic and real-world design methods.

Part 4 deals with the Integrated Method: the new concept of balance for operation (BFO) is described in Chapter 8. Chapter 9 introduces a new approach to using the premises of the physical layout as input data for the logical layout. The concurrent approach to ALD is presented in Chapter 10. Chapter 11 presents the OptiLine software, which is the result of many years of research at the Université Libre de Bruxelles, Belgium, and implemented by the software company Optimal Design (www.optimaldesign.com). Chapter 12 outlines a final summary to assess the significance of what has been covered in this book and suggests new lines for future work.

2 Design Approaches

2.1 Introduction

Designing of manufacturing systems involves the design of products, processes and plant layout before physical construction [35]. CE, which is known as *simultaneous engineering*, allows an interaction among different levels of the design of flexible manufacturing systems. This approach is intended to force the developers and designers, to consider all elements of the product life cycle from conception through disposal, including quality, cost, schedule, and user requirements, *etc. Concurrent engineering is getting the right people together at the right time to identify and resolve design problems* [37]. Figure 2.1 shows the modules composing the CE concept, which can be divided as follows:

Product analysis (PA) is based on classical design for assembly (DFA) rules and proposes a first product design review and a first decomposition of the product into sub-assemblies [18]. It yields a precedence graph between the functional components of the product.

Operating modes and techniques (OMT) proposes an assembly technique (screwing, force fit, *etc.*) for each attachment between the parts, and possible modes (manual, automated, robotic) for each operation [92]. Then, the process time and cost are computed for each chosen technique.

LL assigns tasks to stations and decides on the position of the stations and conveyors.

In this chapter, Section 2.2 explains the difficulty of design, while the design and search approach is presented in Section 2.3. The gap between theory and practice of ALD is discussed in Section 2.4. An approach for the quality of a solution is presented in Section 2.5, and Section 2.6 is devoted to ALD evolution.

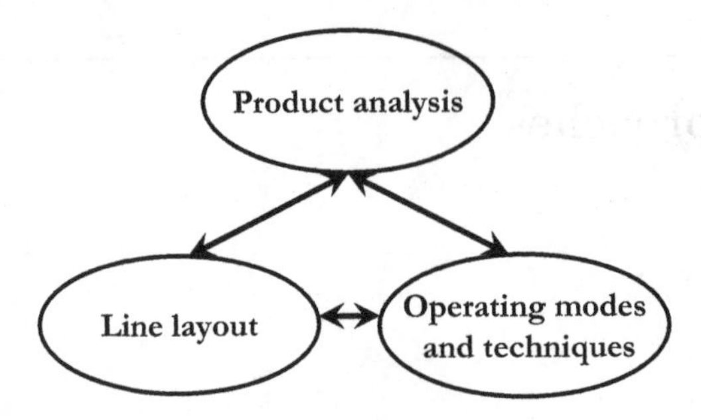

Figure 2.1. Flow chart of the CE

2.2 Why the Design is Difficult?

Design is a prescience phase and it must go through several stages before it constitutes a natural science. In mechanical engineering, a product or a component is evaluated under numerous interrelated criteria, such as quality, reliability, assembly, and maintenance, *etc.* Then, one or more approximate solutions to the problem are selected. Thus, design is very subjective and depends on the background of the designer [106, 108].

2.3 Design and Search Approaches

Design has not always been a rational process; it is often a chaotic affair where consultation and consensus are scarcely evident. The work of participants in the process is often departmentalised, each one with its specific expertise. Participants always explore their ideas unilaterally through virtue of their 'expertise', imposing constraints upon all others. The process begins with the identification and analysis of a problem and proceeds through a structured sequence in which information is researched and ideas are explored and evaluated until the 'optimum' solution to the problem is reached. As we glance through a number of design methods presented in the literature, many circles, arrows, paths, boxes, charts and diagrams can be observed [106, 108].

2.4 The Gap Between Theory and Practice

The operation research community has developed several algorithms to tackle the ALB and RP problems [12, 146]. The adaptation of such algorithms to real-world problems would yield very useful tools, since they are able to propose 'optimal' solutions for benchmarks. Only a few companies use published techniques to balance their ALs because they suffer from substantial loss of information [89, 129, 135]. In fact, little work has been done to model the full range of practical ALD considerations. Generally, we tackle linear ALs without separation into sub-lines. The common performance indices are the cycle time and the number of stations. In fact, other factors (*e.g.* traffic problems, station space, transportation networks, *etc.*) may also heavily affect the system performance. The following sections present some reasons which render the difficulties for academic methods to be applied to real-world problems.

2.4.1 Input Data

Most of the industrial approaches applied to design problems suffer from the amount of data the designer has to introduce. On the other hand, existing academic algorithms require small amounts of input data and cannot be applied to industrial problems [113]. They suffer from substantial loss of information, leading to solving fictitious problems rather than real (industrial) ones. Therefore, there is a clear need to overlap the two concepts and deal with more real constraints of the design problem, rather than spending time on a *benchmarking* fight.

2.4.2 Multiple Objective Problem

The ALD must be formulated as a multiple objective problem rather than minimising the number of stations or the imbalance between stations. Efficient ALD methods should be able to deal with conflicting objectives and consider the user's preferences. They should be quick enough to allow the designer to test as many alternatives as possible (see Chapter 5).

2.4.3 Variability

Most of the ALD parameters that can be accurately estimated by engineers are available in terms of their average values (*e.g.* the mean process time, the average cycle time, and the mean reliability of equipment). In some cases, assigning a fast operator, in the case of manual AL, to the operation with high variability may help to increase line productivity. Stochastic methods must be integrated into ALD approaches to deal with these types of problem (see Chapter 6).

2.4.4 Scheduling

Most research on ALs considers scheduling problems. The ALB and the variant ordering for mixed production have been considered as two separate but related problems. By separating the two problems, sub-optimal solutions are often obtained. Chapter 8 introduces a new concept, called the *BFO* to treat both problems simultaneously.

2.4.5 Layout

The design problem of organising an assembly system into workcentres in a plant is the well-known facilities layout problem. The position of each workcentre determines the costs of transportation and storage. Better solutions can be found by using the premises of the physical layout (PL) as input data for the LL and *vice versa* (see Chapter 9).

2.5 About the Quality of a Design

Performance evaluation generally involves two steps: (1) mathematical model and (2) model solution. Because of the large number of these components, it is difficult to find a simple model to describe a studied system. For this reason, simulation is frequently considered, where the purpose is to develop a mathematical model that resembles as near as possible the real decision situation. Then, a computer is used for solving the problem under various decision circumstances. It is highly important to take into account the operators knowledge (the person who really does the job), about the complexity of the tasks, the grouping of tasks, the process time, and all their experience on all the assembly methods. Thus, interactive and iterative methods have to be developed in order to introduce such knowledge to computer-aided design (CAD) tools. The designers propose a set of AL alternatives, while operators give their experience and criticism of the proposed solutions (Figure 2.2). The aim is to shorten the gap between *frequent talks* about human factors in ALs and the actual reality of things.

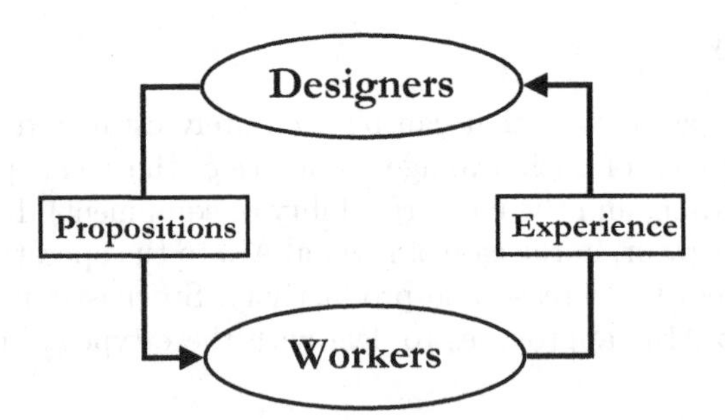

Figure 2.2. Interaction between designers and workers

2.6 Assembly Line Design Evolution

The introduction of new products and the modifications in the product yield frequent redesigns of the AL. Thus, with the increased diversified demand, manufacturers use multi-model ALs. In *batch production*, only one product is produced over a certain period, while in *mixed production* several variants of the product family are produced all the time. In the case of ALD, only a little research has been done on the methods that help to improve existing designs. The aim is to enable a computer to create new designs, with some preliminary or existing designs being supplied. The evolution of complex assembly systems at the same time seems to be more complex, and requires more reflections. As constraints and preferences evolve with time, the progress of design methods has to run parallel to them.

3

Assembly Line: History and Formulation

3.1 Introduction

Assembly work has a long history, and ancient peoples know how to create useful objects composed of multiple parts. However, the objective of modern ALs is to produce high-quality and low-cost products. Manufacturing evolved from single hunter–gatherer to present-day architectures. This chapter starts with the history of the evolution of manufacturing. Next, the ALB problem is introduced. Thereafter, a classification of the ALD problems is given. The question *'why is the balancing problem hard to solve?'* is discussed.

3.2 Evolution of Today's Manufacturing Issues

One of the things that distinguishes humans from the higher primates is the ability to fashion tools to find and process food supplies and make the world a better place to live. The evolution of manufacture has passed through distinct stages at various periods of history [81, 153]. This section examines some stages of the evolutionary cycle of manufacturing and assembly process (see Figure 3.1).

3.2.1 First Metals

From the earliest days, organised manufacturing began to appear. Hunter–gatherers originally formed a simple society that exploited the resources found around them, but they did not attempt to process them. Manufacturing as a specialisation developed first to provide simple hand-tools to hunter–gatherers.

3.2.2 Carpenters and Smiths

As people gathered around the places where surpluses were brought to be traded, towns developed, and this led to the beginning of a middle class. The

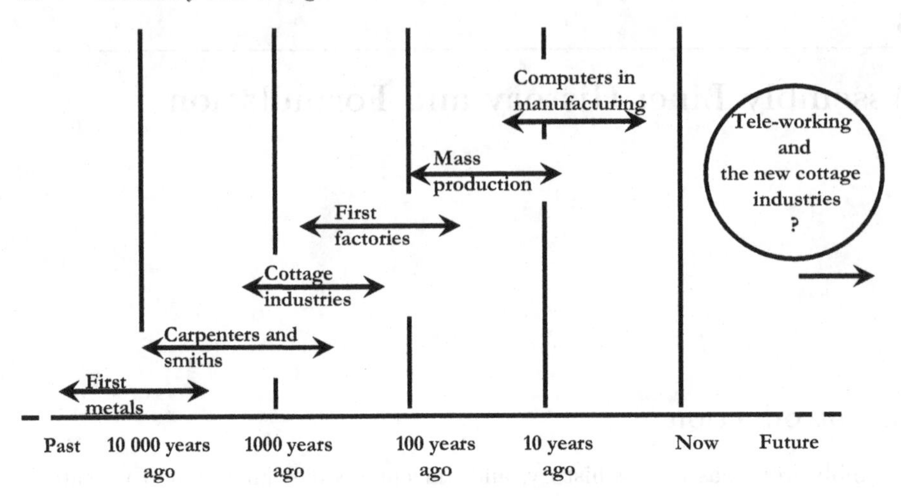

Figure 3.1. Evolution of manufacturing

independent burgesses who traded these surpluses and transporting them from the production place to the use. Manufacturing was concentrated in a number of specialised areas, but it was totally based on skills of individual artisans; (*the person who started to make something, had to finished it*).

3.2.3 Cottage Industries

The next stage in the evolution of manufacturing was the development of cottage industries (artisan or home industry). Then, the cottage industry entrepreneurs saw the opportunities of the factory system. However, it was not the skilled cottage industry workers, but the land-less peasants who eventually went into the factories. Manufacture could have remained with artisans in towns for much longer, but several things happened which precipitated the emergence of the factory system.

3.2.4 Factory System

A factory is a set of people and resources in a place used for manufacturing of products for competitive advantage and superior quality. The rise of factories was dictated by the need for specialisation and concentration. The four prerequisites which were necessary for factories to develop came together in England during the late eighteenth and early nineteenth centuries (Figure 3.2).

3.2.5 Mass Production

The end of the nineteenth century represented an era of good and bad events (mass production and interchangeable parts begun). This was as a conse-

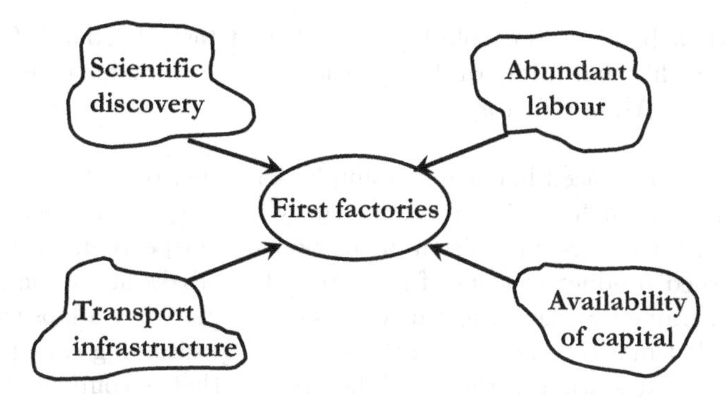

Figure 3.2. Four prerequisites for the development of the factory system

quence of the economic and technological developments. However, mass production focused on managing tiny slices of the work, as managing the whole was not an issue. The emergence of TV advertising increased the demand for mass-produced goods. Customers were introduced to the new world via a tiny screen in their homes. People wanted to choose between *36* different models with a variety of features in different sizes and colours. Manufacturers were confronted with the need to offer a variety of features and to find a way to react quickly to market trends or lose market shares. Mass production and division of labour made it necessary to invent the concept of *work measurement.* Management determined the standard number of hours needed to complete the work and the productivity is measured by man-hours, or by piece-work.

3.2.6 Computers in Manufacturing

The need to superimpose flexibility and variety on a system based on mass production and economies of scale began to cause problems. Confronted with these circumstances, manufactures started to look around to see how they could improve their flexibility and responsiveness. Approaches like flexible manufacturing, just-in-time, and group technology arose at that moment. This speeded up the introduction of computers in manufacturing [81].

3.3 Assembly Line Systems

The automobile industry is a result of a combination of technologies developed by many people over the time. Henry Ford invented the AL, which revolutionised the way cars are made and how much they cost. He was the first to introduce a moving belt in the factory. Employees were able to build cars one piece at a time instead of one car at a time. This principle, which is called 'division of labour', allowed workers to focus on doing one thing very well rather than being responsible for a number of tasks.[1] Charlie Chaplin immortalised in his film 'Modern Times' the way the workers try to repeat themselves on an AL in a factory.

The concept of a paced line is quite simple: a number of stations (four, in the figure: WS1 through WS4) are connected by a conveyor and each station performs one or more tasks (addition of components, inspection, *etc.*) on the partially finished product in front of it (Figure 3.3). Tasks are accomplished by a group of trained workers using machines or robots. After a lapse of time called the cycle time (C), the conveyor moves, thus positioning each product in front of the next station in the line. The product that is completed at the last station leaves the line. Operations are subjected to certain precedence constraints that have predecessors (*i.e.* they can only be performed after one or more other ones have been completed).

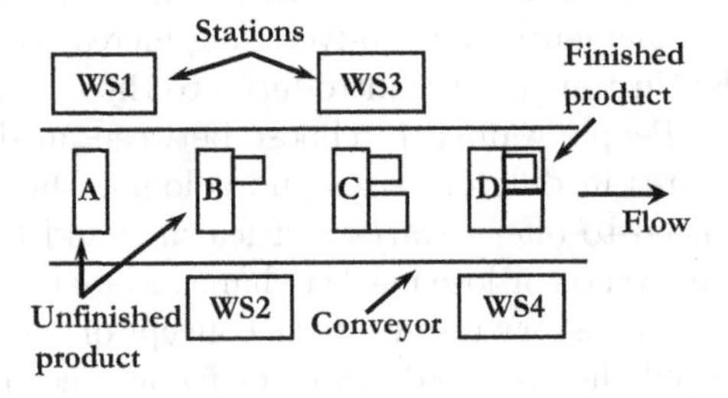

Figure 3.3. AL concept

3.4 Notation and Definitions

In this section, some definitions and notation to describe the ALB problem (ALBP) will be given.

Assembly. This is the process of fitting together various parts in order to create a finished product. Parts may be subdivided into *components* and *sub-assemblies*. The unfinished units of the product are called *work in progress*.

[1] It takes several hours to assemble an automobile; how can a car assembly plant produce a car every few minutes? They use an AL with many stations: each station adds one part or many to the frame in a few minutes and then sends the frame to the next station. After a warm-up period, it may take several hours for one given frame to go through the whole AL, but every few minutes the line produces another car. A few minutes rather than several hours of work: kind of magic...!

AL. This is a flow-line production system composed of a *number of stations* (n) arranged along a conveyor system. The pieces are consecutively launched down the system and are moved from one station to another.

Task. This is a portion of the total work content in an assembly process. The necessary time to perform task is called the *task process time*. Tasks are considered indivisible and they cannot be split into smaller work elements without unnecessary additional work.

Precedence Constraints. These are the order in which tasks must be performed (technological restrictions). The partial ordering of tasks can be illustrated by means of a *precedence graph* [142]. The nodes represent tasks and the directed arcs (i, j) constitute precedence relationships. For example, in Figure 3.4 *task 4* is preceded by *tasks 1* and *2*.

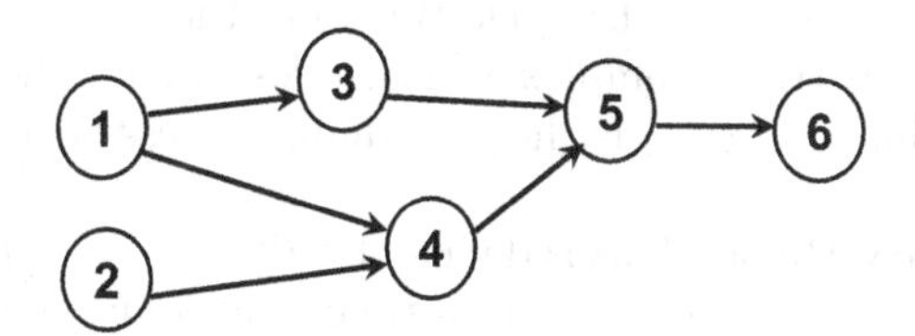

Figure 3.4. Precedence graph

Cycle Time (C). This is the time between the exit of two consecutive products from the line. It represents the maximal amount of work processed by each station. The *desired* C is what the planning department asks for, while the *effective* C (EC) is the real C by which the line will operate.

Capacity Supply (CS). The capacity supply CS $= nC$ is defined as the total time available to assemble each product. The CS is greater or equal to the sum of process time of all tasks' work content. It depends on the transfer system of the AL (free or linked transfer).

Makespan. This is the maximum completion time required to process all operations for a given set of products.

Maximum Peak Time. This is introduced to deal with multi-product ALs and allows some variants process time to lie in the interval $[C, 2C]$. It may not be exceeded by any variant process time on a given station, while the cycle time must not be overstepped by the average working time.

Imbalance. The multi-product ALs are in balance on average. Thus, the process time of each station depends on the variant-product. The imbalance (IB) is measured by the difference between C and the total duration of tasks concerning a given variant on each station (Figure 3.5).

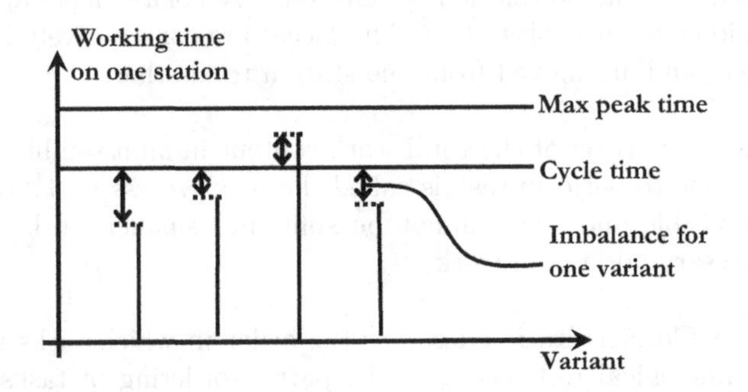

Figure 3.5. Task duration is variable according to the variant

Time Interval. The Time Interval $TI = [t_{min}/C, t_{max}/C] \in [0,1]$ measures the interrelation between the cycle time and the task times. Problems are expected to be relatively complex if TI is close to 1 (where t_{min} is the minimum process time and t_{max} is the maximum process time).

Time Variability Ratio. This is defined by $TVR = t_{max}/t_{min}$ and small values of TVR indicate that the operation times vary only in a small range.

Work Content. This is the sum of process time (T_i) of all tasks:

$$WC = \sum_{i=1..n} T_i$$

Station Time. The work content of a station is referred to as *station load*, the total process time as *Station Time* (ST). The sum of station times of the whole assembly line is the *total assembly time.* The process time of tasks belongs to the interval $[t_{min}, t_{max}]$.

Line Efficiency (E). This is defined by: $E = WC/CS$ measures the capacity utilisation of the line. The unused (idle) capacity is reflected by the *balance delay time*, which is defined by $BD = CS - WC$.

Station Idle Time. This is the positive difference between the cycle time and the station time. The sum of stations idle time is given by $I = CS - WC$. The sum of idle times of all stations is called the *delay time.*

Throughput Time. This denotes the *average* time interval between launching a work-piece down the line and removing the finished product from it.

Smoothness Index (SX). This measures the standard deviation of the distribution of work among the stations.

$$SX = \sqrt{\sum_{i=1}^{i=n}(C - ST(i))^2}$$

3.5 Assembly Line Balancing Problems

Many classifications of AL problems are given in the literature [12, 55, 131, 146]. In this section, some classification schemes of ALBP will be presented.

3.5.1 Assembly Line Models

The demand forecast in assembly and manufacturing may depend generally on the number of product units to be made at each period. Meeting this demand generates requirements for people, equipment, and other resources. Therefore, a demand forecast helps to develop appropriate plans for production. There are three types of production: single product line, family of products line, and multiple products line. The main factors that influence the choice among the three approaches are summarised below.

Single Product AL. The single product line is used to produce only one product. If dynamic phenomena are neglected, the workload of all stations remains constant over time. It is better to use a single AL if the following conditions are true:

- The demand of the product is constant and allows absorption of line cost.
- The product must be delivered in a very short time (no delivery delays).
- The product has a structure which is different from other products.
- The cost of product interdict mistakes during production (made of gold).
- The assembly of the product needs heavy and bulky machines (resources).
- The setup time of the dedicated assembly line (several shifts).

Mixed-Production AL. A family of products is a set of distinguished products (variants), whose main functions are preferably similar. A typical example is a family of cars with different options: some of them will have a sunroof, some will have ABS, *etc.* [153]. These lines can be used in these cases:

- The cycle time is greater than a minute.
- The line price cannot be amortised by a single product.
- The product must not be delivered in a short time.
- Each product is quite similar to others (they constitute a product family).
- The same resources are needed to assemble all the products.
- The setup time of the assembly line needs to be short.

Typically, the products being made on the AL tend to have very similar tasks and precedence diagrams. While designing such ALs, the 'operation stage' must be taken into account.

Batch Production Line. The batch production line is used in the case of multiple different products, or family of products, which presents significant differences in the production processes. Using batch production leads to scheduling and lot-sizing problems. The problem seems to be solved: each product may have its assembly line and the cost will be too high. Thus, designers tend to redesign and reuse existing ALs to produce the different products (batch). The main factors that make the difference compared to the family of products line are:

- The demand of products changes slowly.
- The product must be delivered in a short time.
- The assembly of some products needs heavy and bulky machines.
- The setup time of the assembly line is short.

3.5.2 Variability of Tasks Process Time

The task process time is an essential parameter in the ALB. Simple tasks may have a small process time variance, while complex and unreliable tasks may have highly varying execution times. In the case of human workers, some factors (*e.g.* skills, motivation and communication among the group, *etc.*) have a great influence on the whole AL.

Deterministic Time. In the case of manual ALs, the task time is constant only in the case of highly qualified and motivated workers. More advanced machines and robots are able to work permanently at a constant speed. One can reduce the task time variation by increasing the line's automation degree.

Stochastic Time. In automated flow line, varying production rates may result from machine breakdowns. Furthermore, significant variation may result from non-qualified workers, motivations of the employees, lack of training, *etc.* In the case of a HAL, humans may be hung up by the machine, since a dedicated machine or a robot has the same throughput over time, while human effort varies.

The varying process times of ALs, lead to buffer sizing problems and resources duplication, *etc.* The fuzzy logic[2] concept can be used to tackle the stochastic nature and the variability of process time [174].

[2] Fuzzy logic is a form of knowledge representation suitable for notions that cannot be defined precisely, but which depend upon their context. It enables computerised devices to reason more like humans.

Hidden Times. In the case of automated stations, it is often difficult to determine the operating time of a complex task (two or more grouped tasks). Indeed, the process time of a station is not always the sum of the operating times of each equipment in the group because of the so-called hidden times (see Chapter 7).

Dynamic Time. In the case of human workers, systematic reductions are possible due to the learning effects or successive improvements of the production process. For new tasks, operators take longer times to execute the operation than after becoming familiar with them.

3.5.3 Line Configuration

In the plant layout problem, emphasis is often put on material flow between departments. Ideally, the preliminary analysis of the product and the plant shape leads to a global layout of the AL (several configurations are possible).

Serial Lines. Single stations are arranged in a straight line along a conveying system (Figure 3.6). Each station perform one or more tasks on the partially finished product and can be a simple unit of a complex system [12].

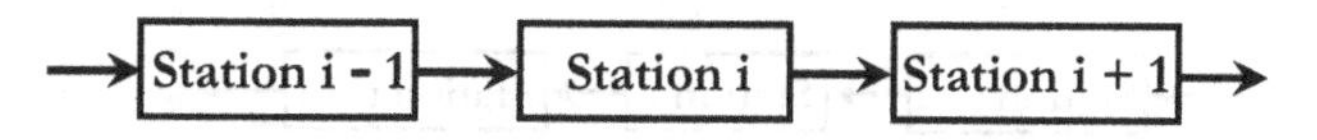

Figure 3.6. Serial line configuration

U-shaped Lines. As a consequence of introducing the JIT production principle, it has been recognised that arranging the stations in a U-line has several advantages over the traditional configuration (see Figure 3.7). Workers are placed in the centre of the 'U' and can monitor each other's progress and collaborate easily whenever required [95]. Thus, workers acquire multiple skills leading to higher motivation, improved quality of products and increased flexibility.

Parallel Stations. With high production rates, the longest task time sometimes exceeds the specified cycle time. A common remedy is to create stations with parallel or serial posts, where two or more workers perform an identical set of tasks. This procedure reduces the average value of the task duration proportionally to the number of workers on the station (see Figure 3.8).

Parallel Lines. It is common to duplicate the entire AL when the demand is high enough. This has the advantage of shortening the AL, but may require more equipment and tooling. Also, if failure occurs at a given station, other lines can continue to run. These units are organised as autonomous work teams (Figure 3.9).

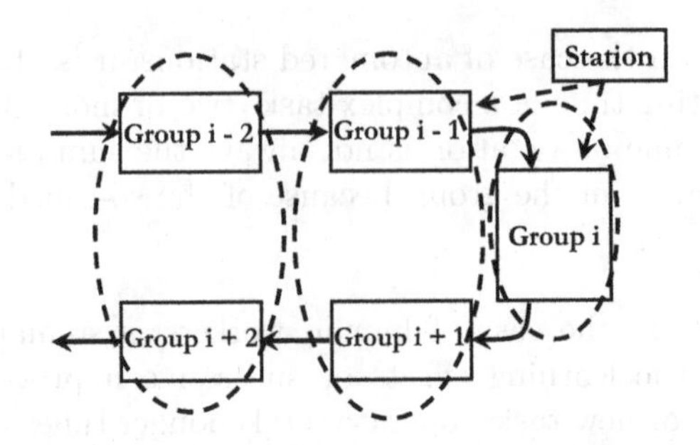

Figure 3.7. U-shaped line configuration

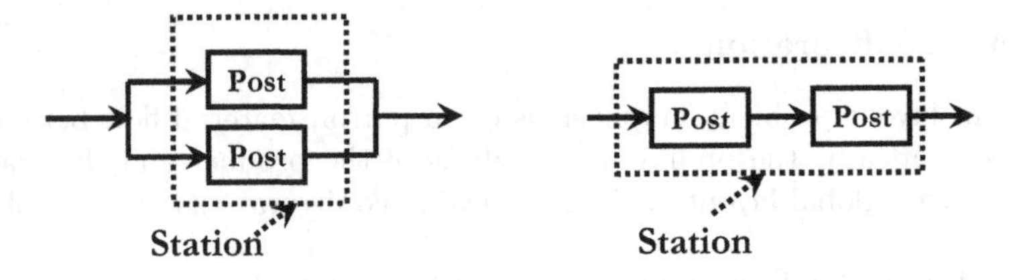

Figure 3.8. Parallel and serial stations

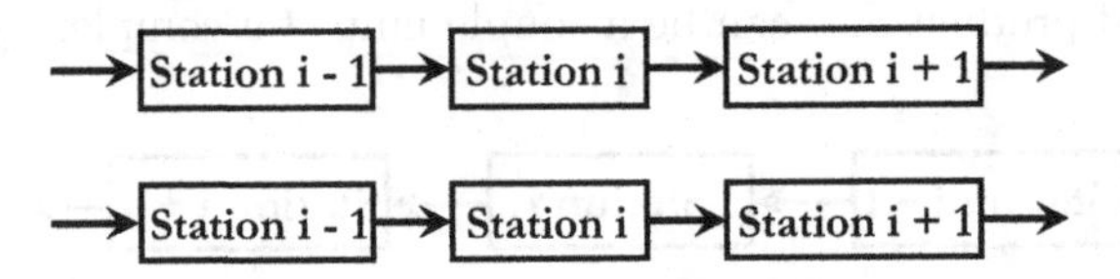

Figure 3.9. Parallel assembly lines

Workcentres. For complex products, the assembly system is most of the time decomposed into sub-systems (workcentres) which are easier to manage than the entire system (see Figure 3.10). The routing of a product between workcentres is fixed, as it works according to a flow topology (see Chapter 9).

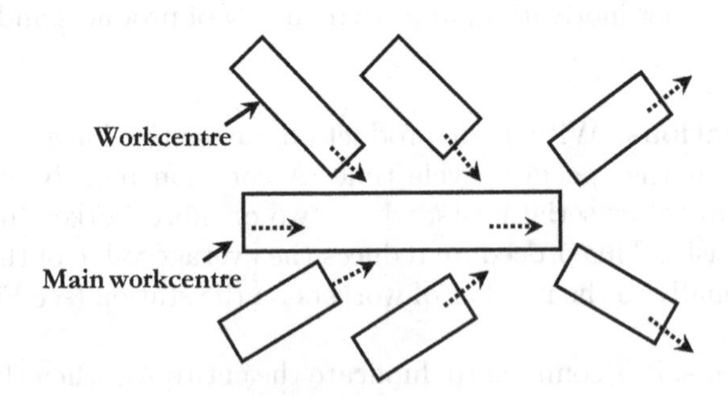

Figure 3.10. Example of plant topology

3.5.4 Additional Constraints

Two types of operation are introduced to deal with these kinds of user preference.

Fixed Operations on Stations. Some operations have to be fixed on a given station (*e.g.* control station, paint station, *etc.*) and no additional operation can be added to it. Figure 3.11 shows two different layouts where the content and the position of the fixed station remain constant. The station occupies the second position.

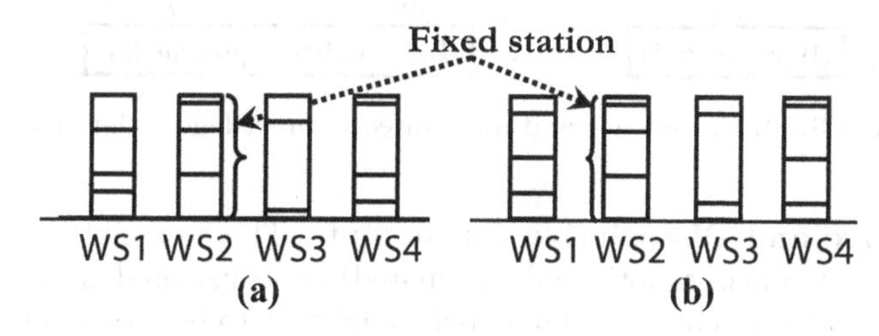

Figure 3.11. Two positions of the fixed station

Linked Operations. A set of operations must be grouped on the same station, but additional operations can be added. In Figure 3.12, a set of linked tasks occupies the fourth position in the first layout, and the third position in the second layout.

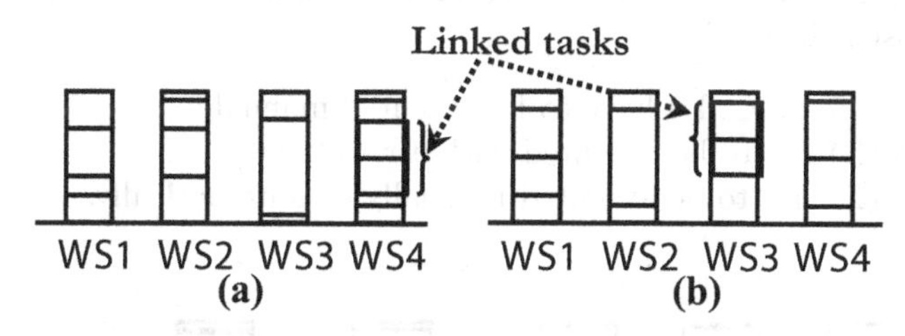

Figure 3.12. Two positions of a set of linked tasks

Must Directly Precede. In order to install safe, non-expensive and stable ALs, designers have to think about the stability of the product being assembled. This leads us to include a *'must directly precede'* relation, which means presence of a direct precedence between tasks rather than an indirect one, as illustrated in Figure 3.13. In Figure 3.13(a), task 1 must only precede task 5. Thus, the acceptable solution consists of the first station which contains task 1, the second tasks {3, 4} and the third one, task 5. Figure 3.13(c) depicts a valid solution where task 1 must directly precede task 5. Indeed, task 1 is directly followed by task 5. Figure 3.13(b) shows a non-valid solution in which stations containing tasks 1 and 5 are separated by a station which contains tasks {3, 4}.

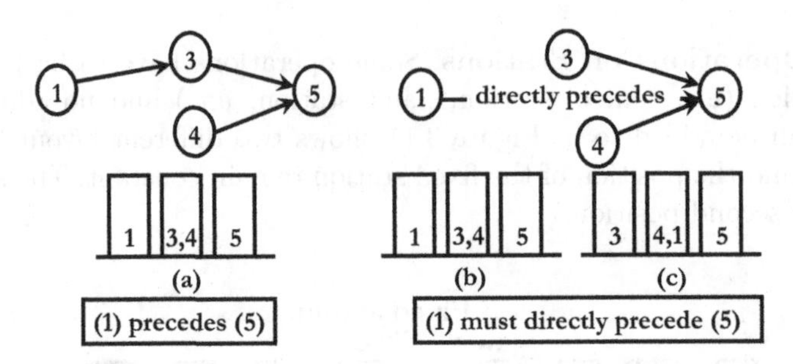

Figure 3.13. Precedence and must-directly precedence relations

Associative and Dissociative Constraints. Three possible modes for each operation (manual, robotic and automated) are considered. This yields a set of *associative* preferences (manual operations have to be grouped together and the robotic or automated operations have to be grouped together) and *dissociative* preferences (manual operations cannot be grouped with robotic or automated operations). The operating mode of tasks will be fixed by the line balancing to obtain the best logical layout. Suppose that, when considering the case where X is manual, the best configuration that can be obtained for the global line is the one represented in Figure 3.14(a) and that case X is automated when the best configuration is the one shown in Figure 3.14(b). In this example, it is clear that the second solution is the best one, if the equipment cost is ignored.

- Operations (A, B, C, D) have to be executed manually.
- Operation (E) has to be executed with a robot.
- Operation (X) has to be executed manually or automatically.

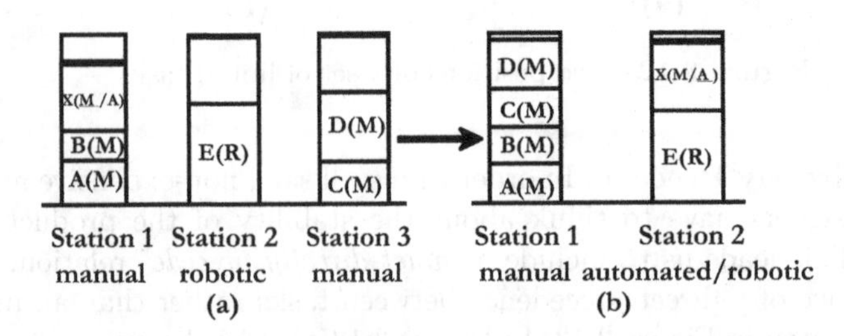

Figure 3.14. Result (a) when X is manual and (b) when X is automated

Positions of Operations. Five positions were introduced to deal with automotive industry: (C: operation on the centre of the car; R: operation on the right side; F: operation on the front; L: operation on the left side; B: operation on the rear). Thus, station duration will be composed of the sum of the operations duration and (1) *operator moves*, and (2) *reading code* of different variants.

3.5.5 Assembly Line Design Problems

The ALD comprises the logical (ALB, RP) and the PL problems. The PL fixes the space requirements taking into account station dimensions and material storage, *etc.* The ordering variants (OVs) allow to consider the operation phase of the line during the design phase.

Line Balancing

The definition of the simple ALB specifies the following assumptions [12]:

1. All the parameters relating to the line must be known with certainty.
2. An operation cannot be divided between two or several stations.
3. Tasks cannot be treated in an arbitrary order due to precedence.
4. All the tasks of an AL must be carried out.
5. All the stations can carry out any task.
6. Process time is independent of the station in which it will be carried out.
7. Any operation can be executed on any station.
8. The AL is serial and does not contain a parallel feeding system.
9. The AL is to be designed for a unique model of a single product.
10. Minimise the number of stations for a fixed cycle time or minimise the cycle time for a fixed number of stations.

Six types of ALBP are proposed. The mathematical formulation of these problems is given below. Let us have a directed acyclic graph $G = (T, P)$ where the nodes T represent the tasks, the arrows P represent the precedence constraints, and a constant L_i (task length) is assigned to each node T_i.

Simple ALBP-1 (ALBP-1). SALBP-1 consists in assigning tasks to stations so that the number of stations is minimised. It is dedicated to a single product with deterministic process times and serial line. Given C (the cycle time) and N as constant, can the T be partitioned into N or less sub-sets S_j (the jth station's tasks) in such a way that: (1) for each of the subsets (the sum of L_i associated with the nodes in the subset does not exceed C); (2) there exists an ordering of the subsets such that whenever two nodes in distinct subsets are jointed by an arrow in G, the arrow goes from a high ordered (earlier) to a lower ordered (later subset)?

SALBP-2. SALBP-2 aims to minimise the sum of the idle times for a given number of stations. It is used in the case of single product, deterministic process times and serial line. Given N stations, can T be partitioned into the N subsets S_j? Since N is fixed, the aim is to minimise the cycle time of the AL.

SALBP-E. SALBP-E aims to minimise the sum of the idle times with a fixed production rate and a fixed number of stations. It is the generalisation of the SALBP-1 and SALBP-2. Given N and C as constant, can T be partitioned into N subsets S_j?. The aim is to find a solution where the sum of the L_i associated with the nodes of the N subset does not exceed C.

Equal Piles for ALBP (EPALP). Given N stations, can T be partitioned into the N subsets S_j? The SALBP-2 minimise the cycle time while EPALP equalise the station loads [125]. The first may lead to unbalanced lines (by minimising the maximal idle time), whereas the second one leads to balanced lines.

ALBP with a fixed number of stations (ALBPF). Given a distribution *frequency* and an *average* duration of each task L_i, assigned to each node T_i, and N stations can T be partitioned into the N subsets S_j?

Multiple Workcentres ALBP (MWkCALBP). Given a set of W directed non-cyclic graphs $G_i = (T_i, P_i)$, a set of N_i, a set of links between these graphs T_i of each graph can be partitioned into the N_i subsets S_{mi}. The aim is to balance a set of workcentres using the different links between them.

Multiple Objective ALBP (MOALBP). AL designers deal with objectives such as line efficiency, smoothness index, and imbalance [27], and the aim is to optimise a set of these objectives.

Resource Planning

Each T_i is characterised by a set of couples $\{L_{i,j}, C_{i,j}\}$ ($L_{i,j}$ is a possible duration of the task and $C_{i,j}$ the cost of the corresponding resource used). Let N, C, and COST be three constants. We define the cost of a subset of T as the sum of $C_{i,j}$ of the nodes belonging to this subset. The question then is how is it possible to find a partition into N subsets of the set of operations and for each of them select a couple $(L_{i,j}, C_{i,j})$ so that, in this case, the sum of $L_{i,j}$s in a partition is less than or equal to C, and the sum of all subsets costs less than or equal to COST?

Ordering Variants

The aim of multi-product ALB is to provide a unique AL valid for all the variants of the product. The variability of the duration of work at each station arises due to the fact that some operations are missing for some variants; stations are balanced on average. The aim is to minimise the imbalance measured by the difference between the average cycle time and the total duration of operations concerning a variant on each station. A large imbalance of the workload among different variants has to be avoided [121, 128, 133]. ALB and OV are interrelated because the balancing solution affects the determination of the launching model.

3.6 Why is the Balancing Problem Hard to Solve?

A special case was considered to show the computational difficulty of the ALB problem using these assumptions: there are no precedence and no grouping preferences. The resulting problem is reduced to packing the tasks into the few number of stations. This is the well-studied 'bin-packing problem' (BPP) which is known as NP-hard problem [171]. Since these problems have been studied for several decades without yielding an easy method, it is widely believed that no such method exists. Therefore, several methods will be used in this book (*e.g.* enumerative methods, evolutionary approaches, heuristics, *etc.*) [123]. As classical methods are quite time consuming for larger problems, our emphasis was on GAs (see Chapters 4 and 5).

Part II

Evolutionary Combinatorial Optimisation

4

Evolutionary Combinatorial Optimisation

4.1 Introduction

There are several reasons for using GAs for design problems. GAs are just one of many methods known in computer science [116]. It is not easy to define exactly which of these methods is best for which problem. However, it is possible, for a given problem, to identify methods that consistently produce better results compared to those produced by other techniques. Rather than spending time, and effort developing new specialised techniques for new problems, most developers prefer to reuse proven algorithms.

Since the 1960s there has been an increasing interest in imitating living beings to develop powerful algorithms for optimisation problems. The term *evolutionary computation* is now in common use to refer to such techniques. Many attempts have been made to understand the adaptive processes of natural systems. Holland [69] attempted to explain the adaptive processes of natural systems. He designed a GA which is an artificial system based upon these natural systems. At the same time, Fogel et al. [48] introduced evolutionary programming, and Schwefel [149] proposed evolution strategies. The first objective of this chapter is to make the reader familiar with the GA technique by introducing its origin and concepts.

4.2 System Organisation

Search algorithms define a design problem in terms of a search problem where the search space is a space filled with a set of points. Each point in that space defines a solution. The design problem is then transformed into the problem of searching for the best solutions somewhere in the space of valid ones. The whole procedure is composed of three steps, which are (1) define the problem, (2) fix the goal and (3) use a method to reach this goal. In tackling a search problem over some space of possible solutions, it is necessary to construct a *representation* of the possible solutions for manipulation and storage. Thus, before applying a GA to any design problem, a certain mapping between human design and the evolutionary method must be made. In order to facilitate the use of GAs, some terminologies must be introduced (*e.g.* gene, allele and locus). The points in the search space are known as *phenotypes*, while their representatives in the solution space are known as *genotypes*. The structures used to represent genotypes are known variously as *genomes* or *chromosomes*. The genotype specifically refers to an individual's genetic structure. The phenotype refers to the observable appearance of an individual (*pheno* in Greek means 'to show'). The process of producing a phenotype from a genotype is known as *morphogenesis* (see Figure 4.1).

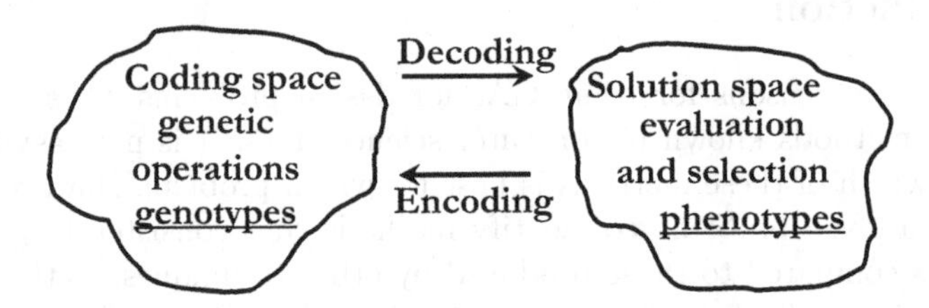

Figure 4.1. Mapping between solution space and search space

Generally, the standard chromosome used to represent a solution typically takes the form of a simple string of values called a *gene*. More formally, a gene can be identified as an equivalence relation over the search space. The particular values that each gene can take are called *alleles*. For example, if the 'eye colour' gene can take values 'blue', 'green' and 'brown' then these are its three possible alleles. The position of a gene in its chromosome is its locus.

4.3 How Do Genetic Algorithms Work?

GAs are a stochastic search technique based on the mechanism of natural selection and natural evolution [57]. All the variations of standard GAs are united by a common thread. The GAs work in parallel with a certain number of chromosomes. The set of individuals (solutions, chromosomes) of each generation is called a *population*. Chromosomes are characterised by their *fitness* and evolve through successive iterations (*generations*). A population of solutions is maintained and the evolution plays the role of adaptation of a population to its environment. This adaptation causes the creation of individuals of increasingly higher 'fitness'. The best solutions are favoured for reproduction every generation and the offspring are then generated from these fit parents using crossover and mutation. Thus, evolution drives the population of better individuals [69].

The standard GA can be summarised by the following steps:

1. The GA can operate on any data type (*representation*) which determines the bounds of the search space. It is desirable that the representation can only encode feasible solutions, so that the objective function (*fitness*) measures only *optimality* and not *feasibility*.
2. The initial population is created during an initialisation phase and it is often generated at random. Generally, some knowledge is used by the GA to start the search from promising regions of the search space.
3. Every member of the population is then evaluated and a fitness value is given according to how well it fulfils the objectives. If there is no clear way to compare the quality of different solutions, then there can be no clear way for the GA to allocate more offspring in the fitter solutions.
4. The GA favours individuals with a higher overall fitness when picking 'parents' from the population. The fitness function allows the evaluation of solutions. Then, these scores are used to determine which individuals will participate in creating the new population.
5. Based on the fitness values, the GA selects candidate solutions and combines (crossover) the best traits of the parents to produce superior children.
6. A small part of the population is mutated. Single existing individuals are modified to produce a single new one. It is more likely to produce harmful or even destructive changes than beneficial ones.
7. Natural *selection* ensures that the weakest creatures die, or at least do not reproduce as successfully as the stronger ones. In the same way, a population is maintained with the fittest solutions being favoured for reproduction. New generations are formed by selecting some parents and offspring and rejecting the less-fit ones.
8. A generation is a population at a particular iteration of the loop. This *iterative* process (selection, crossover, *etc.*) continues until the specified number of generations is passed, or an acceptable solution has emerged.

In the next sections, the existing similarities between natural evolution and GAs are introduced.

4.3.1 Representation

The first step in designing a GA for a particular problem is to devise a suitable representation. For instance, it is quite natural to represent an n-dimensional vector as a string of n values (genes), while it is difficult to represent a graph without introducing extra information (*e.g.* labelling of the nodes).

Encoding

In order to achieve an efficient use of GAs, the encoding must be adapted to the particular search problem at hand. Using a good representation is the first step to narrow the gap between theory and practice in the context of engineering optimisation [30].

Feasibility

GAs may employ four basic strategies to deal with infeasible solutions: rejection, repair, modifying the genetic operator, and assigning penalties [54]. The rejection strategy simply discards all infeasible individuals, while the repairing strategy attempts to create only feasible solutions. For some problems, genetic operators can be modified so that they create only feasible solutions. Finally, penalty functions can be used when infeasible solutions can be recombined to form feasible ones (see Figure 4.2).

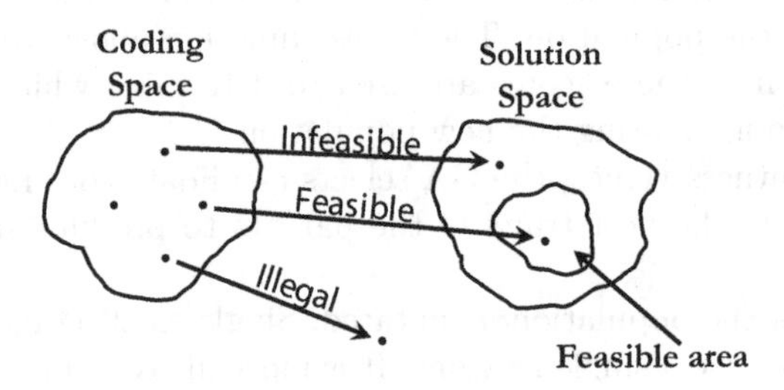

Figure 4.2. Feasibility of solutions

Chromosomes and Solution Spaces

GAs manipulate a coding of the solutions and not the solutions themselves [57]. It is clear that the *1-to-1* mapping is the best used one in which each solution is represented by exactly one chromosome and each chromosome decodes exactly one solution of the original problem. Such an encoding is redundant and its redundancy is a major blow to the efficiency of a GA. The *n-to-1* mapping suffers from a lack of detail because some information is hidden from the GA (see Figure 4.3). For infeasible solutions, many approaches tend to attribute a fitness and an unfitness terms to each genotype. Infeasible solutions tend to have higher unfitness scores [28].

4.3.2 Initialisation of the Population

When initialising GAs with *random* values, the evolution makes extremely rapid progress at first. Indeed, most solutions are largely different and belong to different areas of the search space. Over time, the population begins to converge, with the separate individuals resembling each other more and more [33]. The GA narrows its search in the solution space and reduces the changes

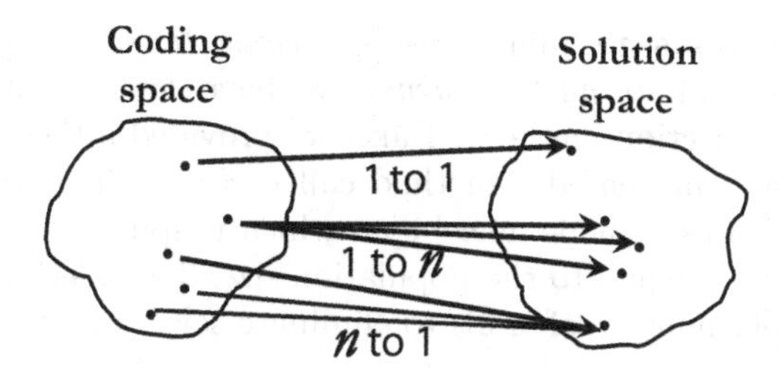

Figure 4.3. Mapping from the encoding to solutions

made by evolution until eventually the population converges to a single solution. The mutation aims to create diversity[1] inside the population.

4.3.3 Sampling Mechanism

Darwin defined *natural selection* or *survival of the fittest* as 'the preservation of favourable individual differences and variations, and the destruction of those that are injurious...' [23]. When selection is the only mechanism at work, the best individual is eventually selected to completely take over the population. The selection mechanism determines which individuals will have all or some of their *genetic material* passed to the next generation. The most commonly used selection schemes are reviewed below.

Roulette Wheel Selection (RWS). the RWS technique works like a roulette wheel in which each slot on the wheel is paired with an individual of the population. The size of each slot is proportional to the corresponding individual fitness. The maximisation problems fit directly into the paradigm *'larger slot implies larger fitness'* [57].

Elitist Models. the first variation is called the *elitist* model and enforces preserving the best solution. The *expected value model* reduces the stochastic errors of the selection mechanism. This is done by introducing a count for each solution s, initially set to the $f(s)/\overline{f}$ value ($f(s)$ is the fitness value of solution s and $\overline{f}$ is the average fitness of the population) and decreased by 0.5 or 1 each time the solution is selected for reproduction with crossover or mutation respectively. Thus, when a chromosome count falls below zero, the solution is no longer available for selection [34].

[1] The process of diversity loss is often the cause of premature convergence, which is the early convergence on an inferior local maximum. A large number of existing techniques are used to maintain diversity in GAs. These include maintaining large population sizes, employing low reproductive or parent-selection pressures, applying mutation, restarting the GA, employing parallel populations, and niche-formation techniques.

Stochastic Methods. techniques like *deterministic sampling*, *remainder stochastic sampling*, and *stochastic tournament* have demonstrated their superiority on simple selection methods. Baker [8] provided a theoretical study of these methods and presented a method called *Stochastic Universal Sampling*. This method uses a single wheel spin which is spun with a number of equally spaced markers equal to the population size. This method gives each individual the proper number of trials to eliminate selection noise.

Tournament Selection. at each iteration the method selects a number k (tournament size) of individuals and selects the best one from this set into the next generation. This process is repeated P times (where P is the population size). It is clear that the large values of k increase the selective pressure of this procedure.

Ranking selection. the solutions are selected proportionally to their *rank* rather than to their evaluation (Pareto optimality). The population is sorted from the best to the worst one, and each individual is copied as many times as possible, and then the proportionate selection is performed. Such methods are more suited for multiple objective problems [49].

The trade-off between exploitation and exploration is generally viewed as one of the key features in an effective search. It is widely accepted that a higher selection pressure leads to *fast convergence*, but also increases the likelihood of premature convergence. On the other hand, very low selection pressure increases the run-time and can causes the *failure to improve* solutions [54].

4.3.4 Genetic Operators

The selection mechanism does not introduce any new solutions for consideration from the search space. It just copies some solutions to form an intermediate population. The second step of the evolution cycle is the recombination which takes the responsibility of introducing new individuals into the population. This is done by the genetic operators: crossover, mutation and inversion. Thus, together, crossover, mutation and inversion allow GAs to discover fit, short and low-order schemata over time.

Crossover

The most popular mechanisim is where two individuals are selected and are crossed over in order to produce offspring. The aim of crossover is to produce new solutions in regions of search space where successful ones have already been found. There are many variations of the crossover operator, and the most common ones are the *'P-point crossover'* and the *'uniform crossover'* [57]. In the P-point crossover, each parent is divided at P locations into P+1 contiguous sections, numbered 1 through P+1. Two offspring are created by exchanging every odd section between the two parents. The uniform crossover can be thought of as P-point crossover, where P+1 is the number of genes in each parent. Therefore, each gene is a section and every section is probabilistically interchanged between the two parents. Other recombination operators take many shapes and forms, including crossover operators that use more than two parents to generate a single offspring. The crossover gives GAs an advantage to perform better than other metaheuristics. Without crossover, GAs lack the additional instruments of the *simulated annealing* (SA)[2] or the *tabu search* (TS) like temperature, tabu list [56, 166].

Mutation

Mutation is a mechanism that has only a small chance of occurring. The standard mutation operator randomly *perturbs* offspring composition by changing a small number of alleles. Unlike crossover, mutation is a unary operator, and only acts on a single individual at a time. Some GAs use only the mutation operator and do not perform any recombination. These GAs are roughly equivalent to running many SA algorithms in parallel and are therefore *mutation-based* methods. Other mutation-based algorithms include the evolution strategy which utilises the deterministic competitions that are always won by the competitor of higher fitness [13]. Mutation maintains *diversity* in the population and can be an answer to the question *'why do children sometimes differ from their parents?'*.

Inversion

Inversion is used to mitigate a *drawback* of the crossover operator. Since the crossing sites are picked at random, longer schemata are disrupted more often than shorter ones. Whenever one of the crossing sites falls between the genes which define the schema,[3] a child will inherit only a part of the schema (unless the other part of the schema is inherited from the other parent). The main drawback is that with the fixed position of loci on the chromosome, it is always the same schemata which are normally subjected to *disruption*. Inversion allows shortening of long schemata by rearranging the positions of loci on the chromosome. In the standard inversion operator, two sites are selected at random and the order of the loci between the sites is reversed.

---

[2] In SA [80], the winner of each tournament is not necessarily the highest fitness individual. This allows for 'stochastic backtracking', where the algorithm can extract itself from dead-ends. The probability of accepting a lower fitness individual over a higher fitness individual can be made to decrease over time.

[3] The schema theorem of Holland [69] shows that the parts of solutions that are observed to perform well (*i.e.* are parts of good solutions) will be sampled with an increasing frequency.

4.4 Landscapes and Fitness

GAs work on a population of candidate solutions using an *objective function*. Applying a fitness function to each one of these chromosomes permits to measure the quality of the solution performance. A fitness *landscape* is a set of points in n-dimension space *(hyper-surface)* obtained by applying the fitness function to every point in the search space. To optimise a function efficiently, the fitness function must be clearly defined and higher fitness individuals must be explicitly promoted. As a result, if any other ability besides propagative success is desired, GAs must directly encourage the formation of individuals with the desired ability.

4.5 Population

A number of population updating modes are used in GAs. The main approaches are the *steady-state* update and *generational* update. A generational update scheme is a population maintenance mechanism in which N children are produced from a population with size N to form the population at the next time-step. This new population of children completely replaces the parent population. In contrast, in the steady-state approach, a single child is produced at each time-step which replaces a single member of the old population. The most straightforward way to maintain population diversity is to *increase the population size*. In large problems, however, restrictions on the computer resources, such as time and memory, make it infeasible to run GAs with the population size needed to maintain the required diversity.

4.6 Simple... but it Works!

The transition rules applied by the GAs occasionally allow the search to 'back-up' to get past local optima and find better solutions. In addition, GAs only use the pay-off information of the objective function to determine which regions to explore, so no other information about the search space is needed. Generally, GAs have two primary tasks, to: (1) *explore* sufficiently a search space in order to locate promising regions, and (2) *exploit* promising regions of the search space in order to focus the search towards the global optimum. Several applications of GAs to real-world COPs lead us to believe that GAs are good optimisation methods. Indeed, the reason why we prefer GAs over other heuristics is due to the concept of combining (*crossover*) parts of good solutions to produce new ones. But there is of course a *catch*. Falkenauer [45] showed that the *encoding* and the *genetic operators* must be adapted to a particular optimisation task. In order to comply with the necessity in practice, we pledge to abandon functions as targets of GA optimisation in profit of optimisation problems [45].

5

Multiple Objective Grouping Genetic Algorithm

5.1 Introduction

Multiple objective optimisation problems (MOOPs) involve two 'quasi inseparable' difficulties: search and the multi-criteria decision aid (MCDA). The space to be searched can be too large and complex to be explored by simple search methods. Section 5.2 outlines the MOOPs, while the related work on MOP is reviewed in Section 5.3. Section 5.4 describes the GGA, while Section 5.5 presents its adaptation to multiple objectives problems. Section 5.6 outlines a case study.

5.2 Multiple Objective Optimisation

In single objective optimisation problems, the feasible set is totally ordered according to the objective function f. For two solutions s_1 and s_2 one has either $f(s_1) > f(s_2)$ or $f(s_1) \leq f(s_2)$. In contrast, mumtiple objective problems (MOPs) present a set of *optimal* solutions which are quite difficult to order. Once, the solutions have evaluated, a vector whose components represent the trade-off in the *decision* search space will be produced. Then a decision maker (DM) implicitly chooses an *acceptable* solution by selecting one of these vectors. The concept of Pareto optimum was formulated by Vilfredo Pareto in 1896 and constitutes the origin of research on MOOP [107]. A solution S_1 is Pareto optimal if there exists no feasible vector S_2 which would decrease some criterion without causing a simultaneous increase in at least one criterion. The two classic strategies that were applied with the traditional separation of search and MCDA can be described as follows:

1. *Make a multi-criteria decision to aggregate objectives, then apply a search method to optimise the resulting figure of merit.* The different objectives are combined to form a scalar objective function, usually through a linear

combination of the attributes. The approach is well suited to proportional non-competing objectives.

2. *Conduct the search using different objectives at the same level of importance.* In the case of such a MOP, a more satisfactory approach is to search for a set of solutions that represents the 'best possible trade-off'. This leads to a set of alternative solutions and the search phase is followed by making a multi-criteria decision to choose among the reduced set. This approach yields the Pareto frontier [127]. Referring to Figure 5.1, O is unique among A, B, C, and D: its corresponding decision vector $O = (O_1, O_2)$ is not dominated by any other decision vector. That means, O is optimal in the sense that it cannot be improved by any objective without causing a degradation in at least one objective. This approach is generally considered to represent a 'best practice'.

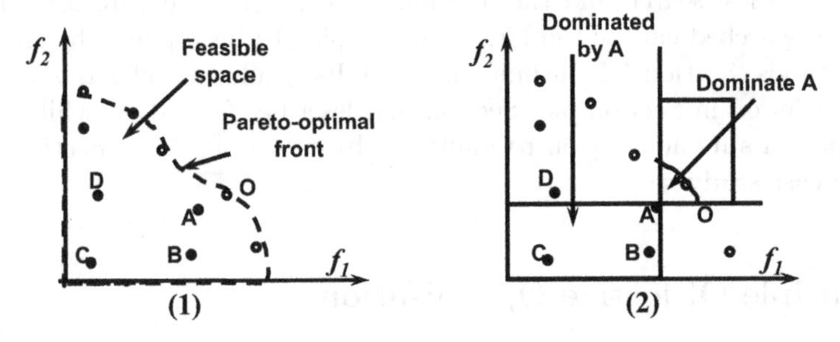

Figure 5.1. Pareto optimality (1) and dominance relations in objective space (2)

Our novel approach is to integrate together decision and search and permit one to deal with user preferences. In the next section, the evolutionary methods for MOPs are described.

5.3 The State of the Art

Rosenberg's [138] study contained a suggestion that would have led to multicriteria optimisation if he had carried it out as presented. While covering the existing literature [38] it seems that the main difference among the methods cited is the way solutions are *ranked.* There are three ranking methods: the *aggregating* approaches, the *non-Pareto* approaches, and the *Pareto* approaches. Three other derivations are the *local search* approach, methods dealing with *preferences* and methods dealing with *constrained search spaces.*

5.3.1 The Use of Aggregating Functions

Many techniques combine several functions in different ways as follows.

Weighted Sum Approach. This approach is used for determining weights when the information about the problem is not enough [49].

Reduction to a Single Objective. The main drawback of this approach is that it is time consuming and the coding of the objective functions may be difficult or even impossible for certain problems [136].

Goal Attainment. In this method, a vector of weights relating the relative *under-* or *over-attainment* of the desired goals must be elicited from the DM in addition to the goal vector [173]. In the case of under-attainment of desired goals, a smaller weighting coefficient is associated with a more important objective. For over-attainment of desired goals, a smaller coefficient is associated with a less important objective.

Use of Penalty Functions. This method is based on both 'constraints satisfaction' method and 'weighting objectives' method. The basic idea is to 'punish' the fitness value of a solution whenever it violates some constraints.

5.3.2 Non-Pareto Approaches

These methods are used to overcome difficulties and the limitations involved in the aggregating approaches.

Vector Evaluated Genetic Algorithm. Schaffer [145] developed an approach approach to use an extension of the simple GA (called vector evaluated GA, 'VEGA'). It was recognised that this would favour solutions with extreme performance in at least one objective.

Lexicographic Ordering. Fourman [50] introduced a lexicographic ordering in which the basic idea is that the designer ranks the objectives in order of importance. In his algorithm, objectives were assigned different priorities by the user and each pair of individuals was compared according to the objective with the highest priority.

Evolutionary Strategies. A multiple objective version is formulated for evolutionary strategies (ESs) [83, 148]. At each step, one of these objectives was selected randomly according to a probability vector and used to delete a fraction of the current population.

Weighted Sum. Hajela and Lin [63] included the weights of each objective in the chromosome and promoted their diversity in the population through fitness sharing. The approach belongs to the category of aggregation selection with parameter variation.

5.3.3 Pareto-based Approaches

In MOPs, there is a set of alternative trade-offs, generally known as Pareto-optimal solutions. In the following, we will review some of the main Pareto-based approaches.

Pareto-based Fitness Assignment. Goldberg [57] suggested the use of non-domination *ranking* and *selection* to move a population toward the Pareto front in a MOP. The author suggested the use of some kind of *niching* (maintain individuals along the non-dominated frontier) to keep GAs from converging to a single point on the front.

Multiple-objective Genetic Algorithm. Fonseca and Fleming [49] proposed a scheme in which the rank of a certain individual corresponds to the number of chromosomes in the current population by which it is dominated. This type of *blocked* fitness assignment is likely to produce a large *selection pressure* that might produce premature convergence. To avoid that, the authors use a *niche-formation* method to distribute the population over the Pareto-optimal region.

Non-Dominated Sorting Genetic Algorithm. Srinivas and Deb [156] proposed the non-dominated sorting GA (NSGA), which is based on several layers of classifications of the individuals. Each individual is assigned a *dummy* fitness value which is proportional to the population size. In order to maintain diversity of the population, these classified individuals are shared with their dummy fitness values. The process continues until all individuals in the population are classified.

Niched Pareto Genetic Algorithm. Horn and Nafpliotis [70] used a Pareto domination tournament (a niched Pareto GA, NPGA) instead of a non-dominated sorting and ranking selection method. Two random individuals are picked to select a winner in a tournament selection. If one of them is non-dominated and the other is dominated, then the non-dominated individual is selected. If both are either non-dominated or dominated, a niche count is found for each individual in the entire population.

5.3.4 Preferences and Local Search Methods

Multi-objective Genetic Local Search. Murata *et al.* [102] proposed a specification method of the local search direction for each solution in the multiple objective genetic local search (MOGLS) algorithm for finding a set of Pareto-optimal solutions. They used a weighted sum of multiple objectives as a fitness function for selecting a pair of parents solutions. The main feature of the selection procedure is that the weights attached to the multiple objective functions are not constant, but randomly specified for each selection.

Local Search. Jaszkiewicz [73] presented a random directions multiple objective genetic local search metaheuristic (RD-MOGLS). This method is based on the idea of simultaneous optimisation of all possible utility functions.

Preferences for Multiple-objective Genetic Algorithm. Cvetković and Parmee [31] presented a method based on preference relations transforming qualitative relationships between objectives into quantitative attributes. This method is integrated in the weighted sum of a GA and a combination of Pareto and a weighted GA. A modified Pareto method for MOPs is presented to yield a Pareto front dealing with preferences.

5.3.5 Constrained Problems

The presence of 'hard' constraints in a MOP may cause extra difficulties. It is sure that the success of a Multiple-objective genetic algorithm (MOGA) in tackling these problems depends on the constraint-handling technique used. Penalty function methods require an appropriate choice of a penalty parameter of each constraint.

Multi-objective Evolution Strategy for Constrained Optimisation Problems. To and Korn [163] presented a multiobjective evolution strategy (MOB-ES) method for solving MOPs subjected to linear and nonlinear constraints. The method is based on the concept of C-(constraints), F-(objective function) and N-(niche) fitness, which allows one to handle constraints and (in)feasible individuals. The authors provided new ideas for handling (in)feasible individuals, since some niche infeasible individuals can be better than some feasible ones.

Strength Pareto Evolutionary Algorithm. Ziztler [175] introduced an evolutionary approach to multi-criteria optimisation, called the strength Pareto evolutionary algorithm (SPEA). The method performs a clustering procedure to reduce the number of individuals externally stored without destroying the characteristics of the trade-off front.

Constraints Handling Through a Multiple-objective Technique. Coello [29] proposed a population-based approach similar to VEGA to handle constraints. The VEGA approach [145] is known to have difficulties in MOPs, since it aims to find individuals that excel only in one dimension regardless of the others. For a comprehensive review, the reader is referred to [29, 38, 49, 167].

5.4 Grouping Problems and the Grouping Genetic Algorithm

Our main objective is to design ALs which are problems of assignment of tasks to stations. These problems can be easily transformed to grouping problems. Falkenauer [44] pointed out the weaknesses of standard GAs when applied to grouping problems and introduced the grouping GA (GGA) to match the structure of grouping problems. The GGA's operators (crossover, mutation and inversion) are group-oriented aimed to follow the structure of grouping problems.

5.4.1 Encoding Scheme

The most distinctive feature of the GGA is the use of a special solution encoding. Falkenauer [44] indicated several drawbacks of standard GAs when applied to grouping problems. These drawbacks, due to the fact that the schemata processed by either the classic GA of Holland [69] or the ordering GA of Goldberg [57], do not represent meaningful regularities of the search space of grouping problems.

Given the fact that schemata processed by a GA are defined over the genes in the chromosomes, it thus follows that a GA adapted for grouping problems must cast groups as genes in its chromosomes. Consequently, the genes in the chromosomes of a GGA encode groups of objects rather than the objects themselves, as illustrated in Figure 5.2. The objects in the upper part of the figure are grouped into six groups of various sizes. Likewise, the corresponding chromosome depicted in the lower part of the figure features six genes and each of them encodes a group of one more objects.

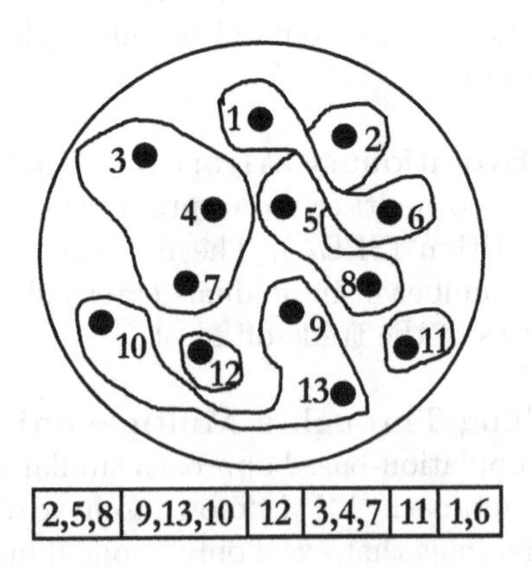

Figure 5.2. A grouping and the corresponding GGA chromosome

5.4.2 Crossover Operator

The GGA crossover proceeds as follows (see Figure 5.3).

1. Select randomly two crossing sites and delimit the crossing section in each of the two parents.
2. Inject the contents of the crossing section of the first parent at the first crossing site of the second parent. Recall that this means injecting some of the groups from the first parent into the second one.
3. Eliminate all objects which occur twice from the groups that they were members in the second parent. In this case the 'old' membership of these objects gives way to the membership specified by the 'new' injected groups. Consequently, some of the 'old' groups coming from the second parent are altered: they do not contain all the objects anymore, since some of those objects had to be eliminated.
4. If necessary, adapt the resulting groups according to hard constraints of the problem and optimise the cost function. At this stage, local problem-dependent heuristics can be applied.
5. Apply the points 2 through 4 to the two parents with their roles reversed in order to generate the second child.

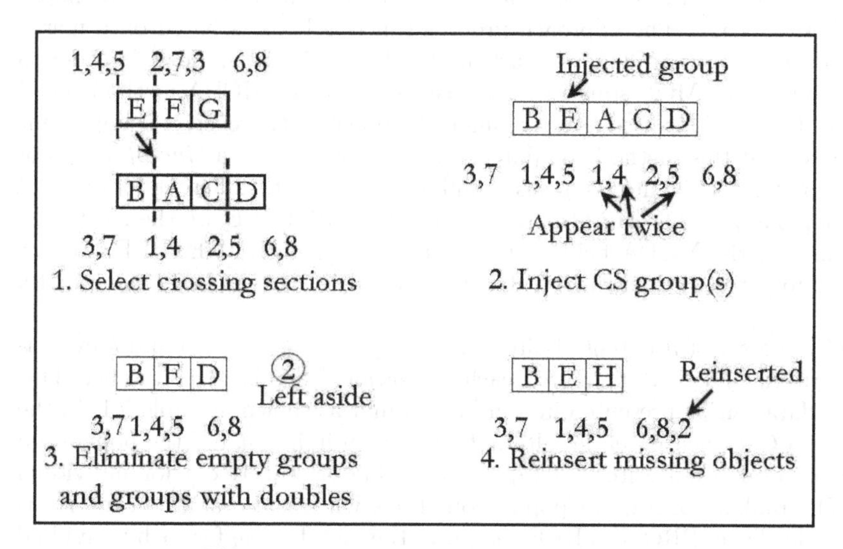

Figure 5.3. The GGA crossover operator

Note that the child inherits genes from the first parent (its crossing section injected in point 1) as well as from the second parent (the genes not affected by the elimination in point 3). Since the genes encode groups, the child inherits groups from both parents, as required by the structure of grouping problems.

5.4.3 Mutation Operator

According to the nature of the particular grouping problem, one or more of the following three operators can be applied: create new group(s) from randomly selected objects; eliminate a randomly selected group by distributing the objects it contains over the other groups; and shuffle a small number of objects among groups.

5.4.4 Inversion Operator

The inversion operator serves to shorten promising schemata made of co-adapted genes. In the GGA, the mechanism is the same as the operator of Holland [69] (*i.e.* a segment on the chromosome is selected at random and the order of genes in that segment is inverted). For more details about the GGA and its applications, the reader is referred to [44].

5.5 Multiple Objective Grouping Genetic Algorithm

Figure 5.4(a) presents the most used approach to MOPs in which a GA generates a set of Pareto solutions and the DM uses his preferences to choose the best solution. The approach proposed in this book is based on a merge of a search and an MCDA, as illustrated in Figure 5.4(b). Indeed, in order to come out of the MOP stated by the *cost function*, the MCDA method called PROMETHEE II is used as a ranking technique [20]. However, it is important to know that this method computes a net flow ϕ which is a *kind of fitness* of each solution. This 'fitness' gives a ranking between the different solutions of the population [122]. ELECTRE [139] and PROMETHEE are the most used methods in the MCDA field. It is preferable to use PROMETHEE because it is simpler to use than ELECTRE and easier to understand and manipulate.

Given a set of n potential alternatives (a_1, a_2, ... a_i, ... a_n) and k evaluation criteria (f_1, f_2, ... f_j, ... f_k), each evaluation $f_j(a_i)$ is a real number. This set of data can be presented in a matrix format as shown in Table 5.1. In the case of a GA, each potential alternative a_i is an individual of the population and the evaluation criteria $f_j(a_i)$ is the value of objective j for individual a_i. The ranking of a given population starts when the evaluation matrix is available. Then, PROMETHEE computes the 'net flow' ϕ_i for each individual i. Weights (associated with each objective) are involved in the computation of the ϕ number and represent the relative influence of each objective. Thus, solutions are not compared according to a cost function yielding an absolute fitness of individuals as in a classical GA, but are compared with each other depending on the current population.

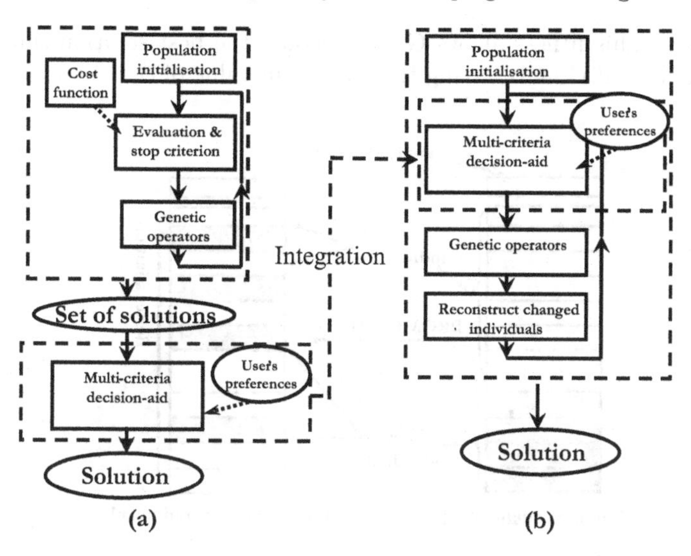

Figure 5.4. Integrating search and decision making into the GA

Table 5.1. PROMETHEE II evaluation matrix

	$f_1(.)$	$f_2(.)$	...	$f_j(.)$	...	$f_k(.)$
a_1	$f_1(a_1)$	$f_2(a_1)$	...	$f_j(a_1)$	...	$f_k(a_1)$
a_2	$f_1(a_2)$	$f_2(a_2)$	...	$f_j(a_2)$	...	$f_k(a_2)$
...						
a_i	$f_1(a_i)$	$f_2(a_i)$	...	$f_j(a_i)$	...	$f_k(a_i)$
...						
a_n	$f_1(a_n)$	$f_2(a_n)$	...	$f_j(a_n)$	...	$f_k(a_n)$
	ϕ_1	ϕ_2		ϕ_j		ϕ_k

In each generation, a ranking changes the fitness of individuals with their *environment.* In classical GAs, the fitness of an individual is independent of the other individuals that constitute the population. There is no direct feedback from the environment to the fitness of the individual, which remains constant and *unaffected* by their environment. This is a weak point of GA-based methods in the case of MOPs. The values of the ϕ are context related and have no absolute meaning. Hence, it becomes impossible to fix a stop criterion for GAs. The optimisation is stopped at the user's request, or if no better solution has been found for a given number of generations.

5.5.1 Control Strategy

In each generation, a set of old and new individuals as well as the best-ever solution are involved in the evaluation of the whole population (Figure 5.5). The MCDA method ranks the individuals taking into account the presence of the others. This fitness allows GAs to choose the best solution simply by looking for the individuals having the maximum value of ϕ.

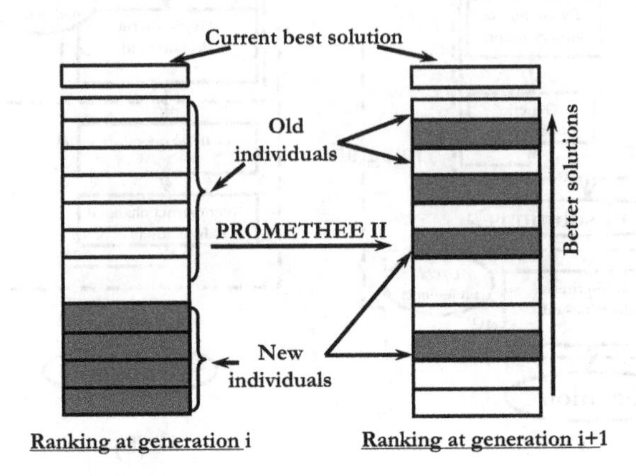

Figure 5.5. Control strategy of the MO-GA

5.5.2 Individual Construction Algorithm

For a given problem, the individual construction algorithm (ICA) is used to create individuals. In the case of an ALB problem, an ICA called 'equal piles' is proposed in Chapter 6, while the 'equal piles/branch and cut' ICA is presented for the RP problem in Chapter 7.

5.5.3 Overall Architecture of the Evolutionary Method

Optimising a combination of the objectives has the advantage of producing a single solution, requiring no further interaction with the DM. For a given user's preferences and a given design problem, the following MOGA is implemented.

The initial population is generated using an ICA. The individuals are then ranked using PROMETHEE II. At each iteration of the main loop, the better solutions are selected from the current population. Recombination produces a number of new individuals. The mutation is used to explore the search space and then the offspring is incorporated into the original population. Again, individuals are ranked using the MCDA and the loop finishes when the termination criteria are satisfied. The basic features of the MO-GGA are presented in the following pseudo-code.

Generate an initial population with an individual construction algorithm; Order individuals using PROMETHEE II;
repeat
Select parents;
Recombine best parents of the population;
Mutate children;
Reconstruct individuals using the ICA;
Replace individuals of the population by children;
Use PROMETHEE II to order the new population;
until *a satisfactory solution has been found.*

5.5.4 Branching on Populations

The method is inspired from the work of Steinberg and Rasheed in their optimisation method by searching a tree of populations [157]. The idea is based on artificial intelligence search techniques like *branching* and *backtracking*, as illustrated in Figure 5.6. Each criterion is attributed a triplet (w, p, q), where:

p is the preference threshold – if the absolute value of the difference between two solutions is higher than p, then this difference is significant and the solution representing the highest performance is better than the other.

q is the indifference threshold – if the absolute value of difference between two solutions is lower than q, then this difference is not significant and the two solutions are practically equivalent.

w is the weight – the weight w means that if a criterion is attributed a weight of 3, and another a weight of 2, that means that two points (respectively three) gained with the first criterion can be compensated by four points (respectively 5.5) gained in the second. It is assumed that the values p and q are the same for the two criteria. Let

- $t_i^g = (p_i^g, q_i^g, w_i^g)$ be the triplet (p, q, w) of objective i at the generation g;
- $T^j = (t_1^j, t_2^j, t_3^j, ..., t_N^j)$ be a set of triplets at generation j;
- N be the number of objectives.

The procedure starts by assigning a triplet to each criterion (T a set of triplets at generation i); the GA is then run for a certain number of iterations g_i. The population obtained after g_i iterations is then analysed by the user (quality of the objectives, the global quality of the best solution, ...). If the user is unsatisfied with the solution, the triplets assigned to the different criterion can be modified. The GA is re-run using the new values of the triplets (T^{i+1}). This technique of branching on populations helps to guide the GA to deal with MOPs looking for the compromise between objectives. A *set of triplets* at generation j is obtained by applying the function M_i^j on the set of triplets at generation i:

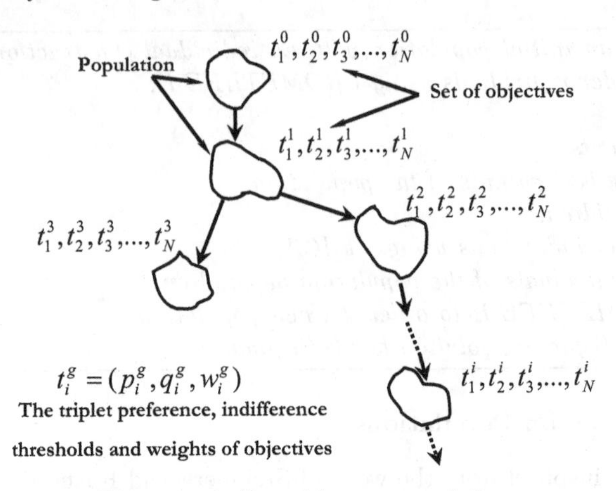

Figure 5.6. Branching on population

$$t_1^j, t_2^j, t_3^j, ..., t_N^j = M_i^j(t_1^i, t_2^i, t_3^i, ..., t_N^i)$$

For instance, suppose we have two objectives and the two triplets at generation 0 are $t_1^0 = (1, 0, 1), t_2^0 = (1, 0, 0)$. Also, by assuming the modifying function M_0^1 which transforms the triplets of generation 0 to those of generation is 1: $t_1^1 = (1.0, 0.5), t_2^1 = (1.0, 0.5)$. Then, the population at generation j is obtained by running the GA and using its corresponding set of triplets $t_1^j, t_2^j, t_3^j, ..., t_N^j$, as illustrated in Figure 5.7.

$$P^j = M_i^j(P)$$

where:

- M_i^j is the function for modifying triplets of generation i to triplets of generation j,
- P^i is the population at generation i,
- $M_i^j(P^i)$ transforms population of generation i using triplets $t_1^j, t_2^j, t_3^j, ..., t_N^j$.

$T_1 = t_1^1, t_2^1, t_3^1, ..., t_N^1$ $\quad$ $T_2 = t_1^2, t_2^2, t_3^2, ..., t_N^2$

P1 $\rightarrow$ P2

$T_2 = M_1^2(T_1)$

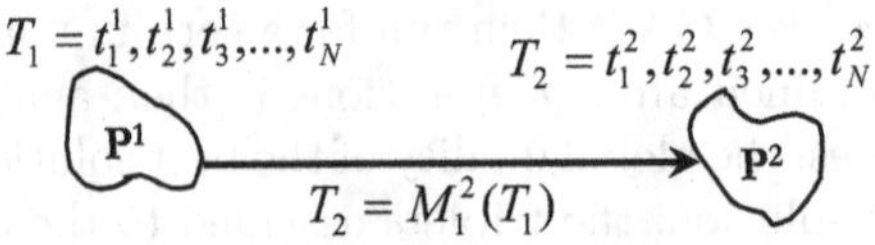

Figure 5.7. Evolving from generation 1 to generation 2

The term *generation* refers to the order of the modification which is set by the DM after analysing the quality of the best solution for the given set of triplets. The whole process of optimisation is composed of a set of R runs of GAs using different sets of triplets. If the results of a modification cannot be accepted due to an inappropriate setting of the triplets, the DM has the choice between starting the run from the beginning or returning to the last triplets and restarting from the last population.

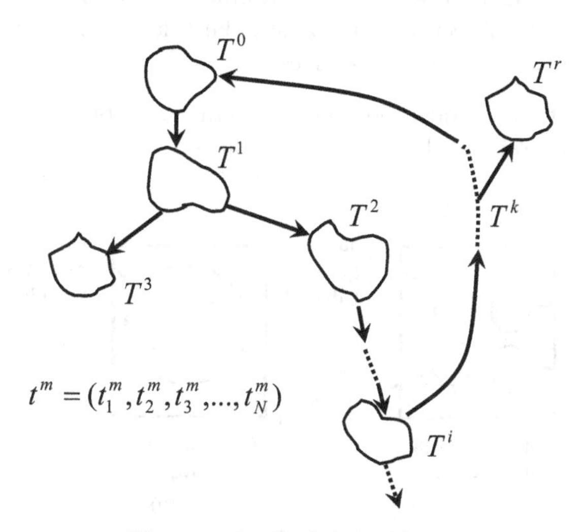

Figure 5.8. Cycling approach

The modification function must be *reversible* (*i.e.* the given set of triplets yield the same population regardless of the stochastic behaviour of GAs). Abstraction is done from the initial starting population (*i.e.* results obtained using a given set of triplets are independent of the starting population) as illustrated in Figure 5.8. The population obtained using the triplets T^0 must be close to the population obtained using the sequence of triplets T^0 T^1 T^2 ...T^i... T^k... T^0. The set of triplets allows one to guide the GA during the search phase.

The branching factor (the number of changes of a given number of triplets) in the case of k objectives is $N = 2^{3k} - 1$ (64 in the case of *two* objectives). It is the number of different groups composed of n objects (the factor 3 is due to the fact that for each objective we have a triplet p, q, w). These changes are not always possible and are far from realistic. In the next section, a detailed example is introduced to demonstrate the above concepts.

5.6 The Detailed Example

A set of N objects have to be grouped in a set of groups. Let $Objs_i = \{i/i \in [0..N]\}$ be the set of objects in group i and the size of each object is equal to its identity. A graphical representation of the problem is given in Figure 5.9. In the first solution each object 1, 2, 3, 4 is put in its own group, while in the second solution all objects are grouped in the same group. Each group is characterised by its average and standard deviation. The average of a given group is the sum of the ID of its objects divided by the number of its items (size). For instance, in Figure 5.9(a) the average of group 4 is 4 since it is composed only by the object 4, while its standard deviation is null. The aim is to optimise the two following objectives:

- minimise the standard deviation on the average of groups;
- minimise the average of the standard deviations.

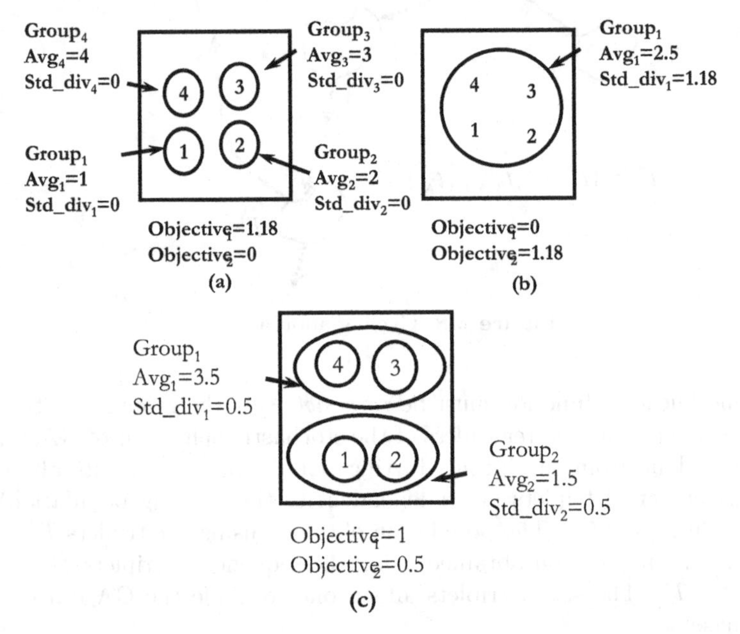

Figure 5.9. One object by group (a) and one group for all objects (b) and two objects by group (c) solutions

Let NbGrp be the number of groups of a given solution, and $size_i$ is the size of each group. The average of a group i is the sum of the size of objects of the group divided by its size and is given by

$$Avg_i = \frac{\sum_j^{size_i} Objs_i[j]}{size_i}$$

The average of a given solution is the sum of the average of groups divided by the number of groups and is given by

$$AVG = \frac{\sum_{i=0}^{NbGrp} Avg_i}{NbGrp}$$

The standard deviation of a group indicates how closely the size of objects are clustered around the average:

$$Std\_div_i = \sqrt{\frac{\sum_{i=0}^{size_i}(Objs_i[j] - Avg_i)^2}{size_i}}$$

The first objective is the standard deviation of the standard deviation of the groups:

$$Minimize : objective1 = \sqrt{\frac{\sum_{i=0}^{NbGrp}(Avg_i - AVG)^2}{NbGrp}}$$

The second objective is the average of the different standard deviation:

$$Minimize : objective2 = \frac{\sum_{i=0}^{NbGrp}(Std\_div_i)^2}{NbGrp}$$

The formulation of the two objectives shows that (see Figure 5.9):

- The first objective takes the value '0' if there is only one group and at the same time the second objective takes its maximal value.
- The second objective takes the value '0' if there only one object by group while the first objective takes its maximal value.[1]

Figure 5.10 presents the evolution of the two objectives where the second objective is neglected (w1=1.0, w2=0.0). The word 'neglected' means that the weight or the preference attributed to the given objective is set to zero. This figure shows that the method tends to minimise objective 1 and ignores the other.

Figure 5.11 represents the evolution of the two objectives where objective 1 is neglected. The weight attributed to objective 1 is null (w1=0.0, w2=1.0). The figure shows that the method tends to minimise objective 2 and pays less attention to the evolution of objective 1. When the weight of an objective is set to zero, the MO-GGA will ignore the importance of this objective. Thus, minimising objective 2 tends to maximise objective 1 and *vice versa*. When optimising the two objectives, a simple way is to set the weights to (w1=0.5, w2=0.5). Figure 5.12 shows that after few generations the two objectives conflict with each other (*i.e.* minimising one objective tends to maximise the other).

[1] Note that a solution composed of two groups where the first group contains objects {1, 4} and the second {2, 3} takes the value '0' for the first objective and the value '1' for the second. This solution is not optimal in the case where the aim is to find the minimum value of objective 1 and the maximal value of objective 2.

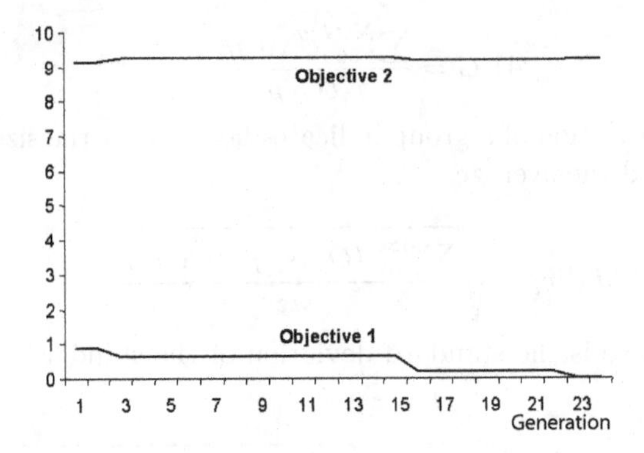

Figure 5.10. Evolution of objective 1 in the case that objective 2 is neglected

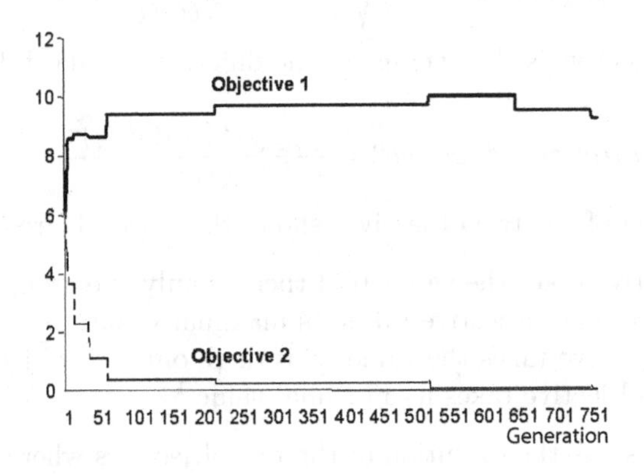

Figure 5.11. Evolution of objective 2 in the case that objective 1 is neglected

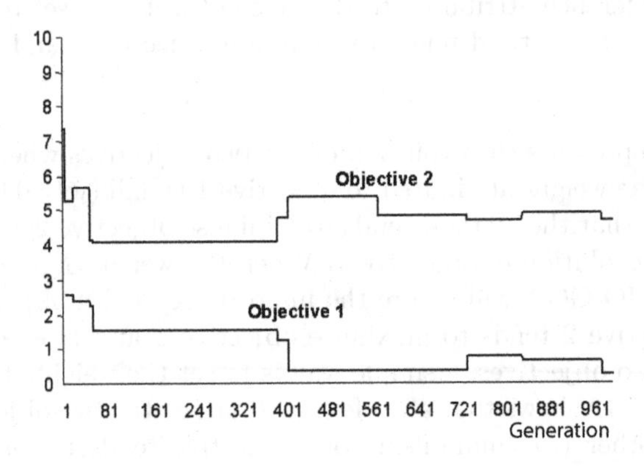

Figure 5.12. Evolution of two objectives having the same preference

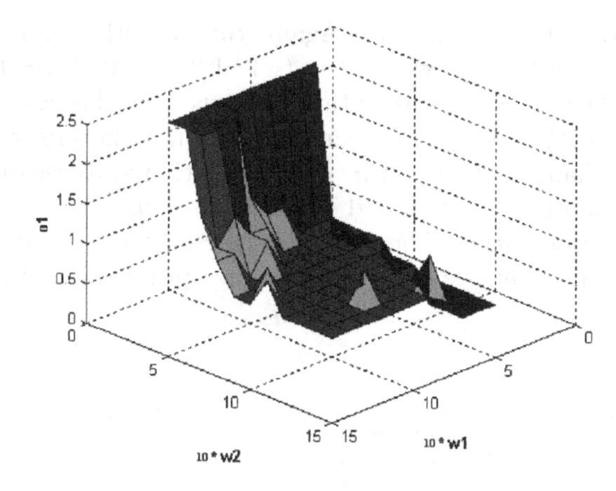

Figure 5.13. Evolution of objective 1 with preferences of the two objectives

The evolution of objectives 1 and 2 for different values of w1 and w2 (as shown in Figure 5.13 and Figure 5.14 respectively) shows that the weights given to the different objectives can guide the algorithm. The figures show that, for small values of w1, the algorithm tends to optimise objective 2 and for small values of w2 the algorithm tends to optimise objective 1. For similar values of w1 and w2, the algorithm simultaneously optimises the two objectives.

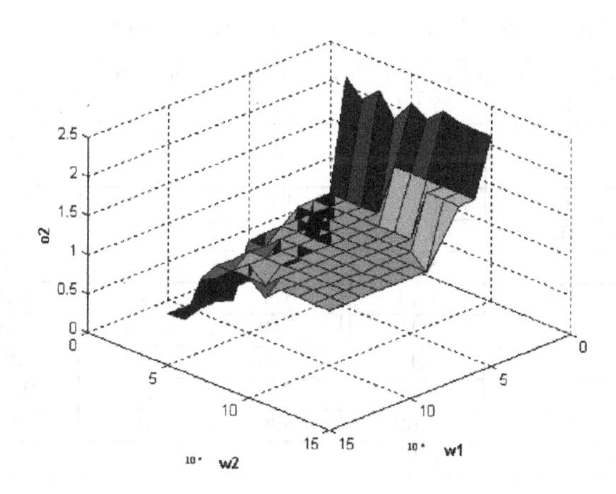

Figure 5.14. Evolution of objective 2 with preferences of the two objectives

The following test was done to verify the idea of branching on population. For the same problem, we started with the given preferences (w1=1.0, w2=0.0). The algorithm was stopped after 180 generations, then the preferences were set to (w1=0.0, w2=1.0). Again, the algorithm was stopped at generation 570, the preferences were set to (w1=1.0, w2=0.0) (see Figure 5.15 and Table 5.2). The graphic representation shows that the best solution of the population and the search direction changes once the preferences given to the objectives change. The fact that the population switches quickly from one direction to another is attributed to the *mutation*, which makes a kind of *diversity* in the population. These results show that the proposed method satisfies the user's preferences with regard to the optimisation objective.

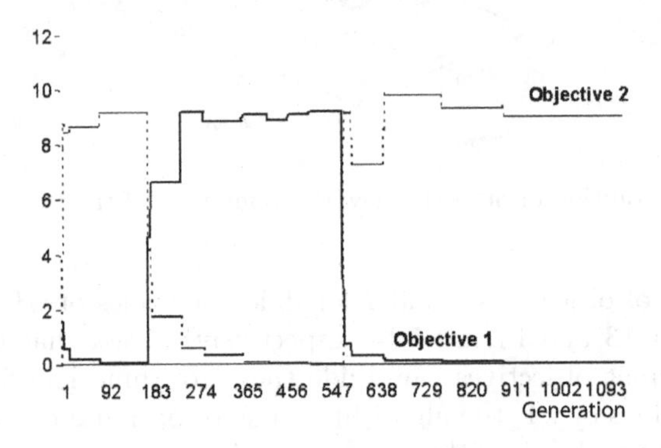

Figure 5.15. Changing the search direction by changing preferences

Table 5.2. Critical values of the two objectives

generation	objective 1	objective 2	generation	objective 1	objective 2
2	1.59	7.92	291	8.87	0.38
3	1.59	7.92	368	9.1	0.11
4	1.02	8.78	420	8.89	0.06
5	1.02	8.78	462	9.11	0.03
6	0.56	8.51	504	9.23	0
18	0.22	8.71	570	5.4	7.27
78	0.06	9.21	572	0.75	9.19
177	4.67	4.19	589	0.32	7.3
183	4.67	4.19	655	0.12	9.81
184	6.67	1.75	772	0.07	9.35
245	9.21	0.62	896	0	9.02
290	9.21	0.62			

Part III

Assembly Line Layout

6

Equal Piles for Assembly Line Balancing

6.1 Introduction

The assembly activities performed within the assembly system not only determine the final qualities of the products, but also affect time-to-market, delivery, *etc.* For manual ALs the most interesting performance index is station workloads' balancing. In this chapter we focus on single and multi-problem ALB (MPALB) problems. Section 6.2 is dedicated to the ALB, and Section 6.3 describes the proposed method of 'equal piles' for ALs. Section 6.4 introduces the MPALB problem.

6.2 The State of the Art

Related AL design (ALD) problems and issues are characterised in the literature as the ALB problem, which usually refers to a single product AL. Most of this literature deals with the maximisation of the line efficiency through minimisation of idle time. In the following, a review of existing methods is presented.

6.2.1 Exact Methods

Several approaches for determining *lower bounds* on the number of stations (n) in the case of SALBP-1 (the cycle time in the case of SALBP-2) are proposed in the literature. The lower bounds are obtained by solving problems which are derived from the considered problem by omitting or relaxing constraints. Most of these techniques fall into two categories, which are dynamic programming and branch and bound methods. A good introduction to optimal approaches of the ALB problem can be found in [12], while a good survey on exact methods for the ALB problem can be found in [146].

Dynamic Programming

The dynamic programming (DP) method is applied to the most COPs and involves the optimisation of multi-stage decision procedures. A given problem is divided into sub-problems which are sequentially solved until the initial problem is finally solved. States at a particular stage s are transformed to states at the subsequent stage $s+1$ by a decision. The generation of states is described by *transformation functions* which depend on the current state and the decision taken. A sequence of decisions, which transforms a state at a stage s to a stage $s' > s$, is called *policy.* DP is a solving approach rather than a technique, and the following approaches are linked to this technique.

Salveson [143] empirically addressed the ALB problem by formulating SALB as a linear programming (LP) problem including all possible combination of station assignments. Bowman [19] provided a 'non-divisibility' constraint by changing the LP formulation to zero–one integer programming. Patterson and Albracht [109] used the integer programming search technique which was efficiently computational more than some of the earlier ones. Jackson [72] proposed an algorithm for SALBP-1 using the notion of a tree. The SALBP was represented by a *tree* in which each path corresponds to a feasible solution with each arc representing a station. The method starts by generating all feasible assignments to the first station (one of these feasible assignments will obviously be part of the optimal solution). Then it generates all feasible assignments to the second station and giving the first station assignments. Then, for each first–second station combination, all feasible solutions are constructed for the third station, and this process is repeated each time by adding one station. For more information about DP, the reader is referred to [1, 62, 111, 151].

Branch and Bound

The branch and bound (B&B) algorithm consists of two main components: the *branching* and the *bounding.* The initial solution is developed into several sub-problems (branching). A multi-level enumeration is constructed by continuously developing such sub-problems for which the optimal solution is already known and need not be branched. These sub-problems are referred to as *leaf* nodes. A path from the root node to any other node of the tree is called a *branch.* Bounding is applied to reduce the size of the enumeration trees. This is achieved by computing lower bounds (the cycle time in the case of SALBP-2) at least necessary for a feasible solution in each node. An optimal solution is found if the 'global' lower bound is found. For more information about B&B, the reader is referred to [16, 68, 74, 146, 147, 171].

Graph Search Technique

Johnson [74] proposed a depth-first-search method called 'fast algorithm for balancing line effectively' (FABLE). Sub-problems are constructed by adding an assignable task to the currently considered station k (starting with station 1). If no such tasks exists, the current station load is maximal and the consecutive station $k + 1$ is opened. In each of the n iterations ($i = 1, ..n$), one non-marked task with the largest process time (which has no predecessor or only marked predecessors) gets the number i and is marked. Whenever a station is opened, the task with the smallest number among the assignable tasks is added. Any further tasks in the station must have a larger number than the task assigned in the ancestor node. Then, the current branch is traced back by removing tasks assignments until an alternative branch can be followed.

6.2.2 Approximated Methods

Talbot *et al.* [161] divided heuristics for ALBP into four categories: single-pass, composite, backtracking and time trapped optimising approaches. In this section, some well-known heuristics will be considered. These methods are divided into simple heuristics and metaheuristics.

Simple Heuristic Methods

One of the first proposed heuristics was the ranked positional weight (RPW) [65]. The main idea is to assign the tasks which have long chains of succeeding tasks. The length of the chain can be measured either by the number of successor tasks or the sum of the task times of the successor tasks. The sum of the task process time and the process times of the successor tasks is defined as the *positional weight* of the task. The tasks are ranked in descending order of the positional weights with arbitrarily broken ties. The tasks are then picked up in their ranked order and assigned to stations if: (1) all predecessors of the task have already been assigned, and (2) the task fits in the remaining time on the station. If a task does not fit in the remaining time on a station, it is skipped and the next task in the ranked order is selected. If no task fits in the station, the station is closed and a new station is opened. The 'reversed RPW' (RRPW) is the RPW method applied to the problem with the reversed precedence constraints. After a balance is found for the reversed problem, the problem is 'un-reversed' to get a balance solution for the original problem [65].

On the side of interactive and iterative methods (balancing and simulation), Praça and Ramos [118] proposed an architecture based on multi-agent simulation called (SimBa). The *balancing* module, which implements three heuristic rules from [17] and [65], allows one to obtain different line configurations, while the *simulation* module evaluates the performance of the proposed configurations. Many heuristics proposed in literature are a combination of the following methods [161]: greater number of immediate successors, greater number of successors, smallest upper bound, smallest upper bound divided by the number of successors, greatest processing time divided by the upper bound, smallest lower bound, minimum slack time, and minimum number of successors divided by task slack, *etc.* For more information about these heuristics, the reader is referred to [6, 17, 52, 67, 68, 78, 101, 119, 150, 154].

Metaheuristics

Garey and Johnson [53] proposed the simple 'first fit descending' (FFD) heuristic for the BPP. The idea is to start with one empty bin, take the items one by one (classified in decreasing size order) and, for each of them, first to search the bins that have so far been used for a space large enough to accommodate the item. If such a bin can be found, put the item there; if not, request a new bin. Putting the item into the first available bin found yields the 'first fit' (FF) heuristic. Searching for the most filled bin still having enough space for the item yields the 'best fit' (BF) heuristic which can perform as well (as bad) as the FF, while being slower. Fitted with acyclic precedence constraints [142], the BPP becomes the ALBP. A modification of the FFD heuristic (augmenting the size of objects with the size of all their predecessors) yields a simple heuristic whose its performance is similar to the FFD heuristic [27].

On the evolutionary algorithm side, to the best of our knowledge, the first attempts were by Falkenauer and Delchambre [46], who used a GGA. They generalised their bin-packing GGA to obtain a fast algorithm supplying high-quality approximated solutions of the ALBP. One advantage of their method was the ability to handle problems with sparse, even empty precedence constraints. Note that the mechanism of complying with precedence constraints proposed in [46] applies equally well to the hybrid GGA of [43], leading to an equally powerful algorithm for SALBP-1. For more information about metaheuristics, the reader is referred to [4, 6, 24, 52, 56, 79, 80, 85, 88, 94, 99, 101, 112, 117, 141, 159, 160, 161].

6.3 Equal Piles for Assembly Line Balancing

When focusing on cycle time, an imbalanced allocation of tasks among stations will typically occur [119]. Thus, using the word *balancing* when cycle time is regarded as a constraint is not appropriate. Indeed, SALBP-1 or SALBP-2 methods do not minimise the imbalance between stations. When dealing with manual multi-product ALs in a batch-oriented approach, most of the time the designer tends to reuse and rebalance the same line for the different batchs. As the allocated space to the AL (number of stations, heavy machines, *etc.*) is generally fixed, designers have to deal with these constraints. This also may occur when the designer has to reuse existing stations to assemble new products. We propose the equal piles approach to balance ALs which seeks to assign tasks to a fixed number of stations in such a way that the workload of each station is nearly equal.

6.3.1 Motivation and Inspiration From Nature

The simple equal piles problem (which does not take precedence constraints into account) was first defined by Jones and Betramo [76]. This problem can be introduced as follows: given a set of N objects of various sizes (one dimension), distribute the objects into K 'piles' in such a way that the 'heights' of the piles are as equal as possible. To solve this problem, Falkenauer [42] presented a GGA combined with the FFD heuristic. Having initialised the required number of groups at random, the remaining items are sorted in decreasing size order and successively added to the 'smallest pile'.

The method presented in this book is based on observing the boundary stones along the motorway (Figure 6.1). Indeed, the distance between two cities is fixed by the number of boundaries (which is equivalent to the number of kilometres). In the case of a flat motorway and fixed speed, the time to cover the distance between two adjacent boundaries is roughly the same (well balanced). It is known that, in order to reduce the time to cover the distance between two cities, the speed must be increased. Moreover, to have an average speed equal to the theoretical one, the safe way is (1) to get back missed time between two boundaries by speeding up between others, or (2) to lose time earned between two boundaries by reducing the speed between others.

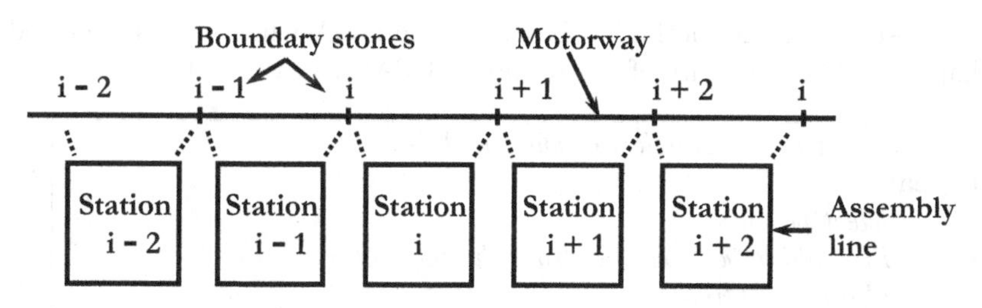

Figure 6.1. Correspondence between an assembly line and a motorway

In the case of an AL, the space is fixed by the designer and the speed of the line depends on the desired throughput of the line and the process time of the tasks. In this case, the distance between two boundaries is equivalent to the process time of a station. Thus, the product has to cross the line in a duration fixed by the desired cycle time and the number of stations. Moreover, to keep the real throughput as close as possible to the desired one, a practical way is (1) to get back missed process time on a given station by speeding up the others, or (2) to lose time earned on a given station by reducing the speed of the others. The hard constraint is no longer the cycle time, but the number of stations.

6.3.2 Input Data

The approach needs the following input data as illustrated in Figure 6.2:

- the desired number of stations;
- the duration of each operation;
- the assembly plan (precedence graph) of the product;
- the user's preference constraints (associative and dissociative).

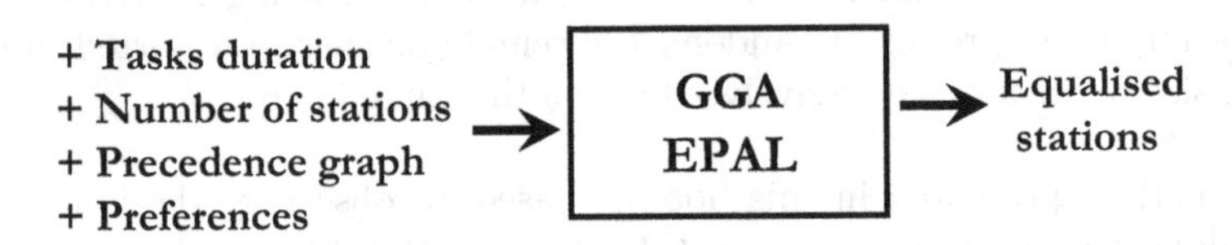

Figure 6.2. Data flow for EPAL

The two first input data are used to estimate the desired cycle time of the AL, while the others allow one to deal with the user's needs (sequence of operations, separate some operations and group some others, *etc.*).

6.3.3 Customising the Grouping Genetic Algorithm to the Equal Piles Assembly Line Problem

The proposed heuristic method is based on the GGA which was introduced in Chapter 5. The structure of the proposed EPAL-GGA is as follows:

```
Generate an initial population with an ICA;
repeat
    Select parents;
    Recombine best parents from the population;
    Mutate children;
    Reconstruct individuals using the ICA;
    Replace individuals of the population by children;
until a satisfactory solution has been found.
```

The purpose of the ICA (EPAL) heuristic is to allocate tasks to a fixed number of stations. The EPAL approach is embedded in the GGA and is used to construct individuals of the population at each generation. The essential and distinct concepts adopted by the method will be described below, along with step-by-step execution procedure and an illustrated example.

Boundary Stones Algorithm

The proposed approach is based on boundary stones. These boundaries will be used as seeds to fill stations and the following steps are used:

Step 1. This begins by detecting if the precedence graph is cyclic [151].

Step 2. This orders the operations using the labels defined below. The labels of tasks depend on the numbers of their predecessors and successors using

$$label1_i = nbpreds_i - nbsuccs_i \tag{6.1}$$

where $label1_i$ is the ordering criteria of operation i (it depends heavily on the precedence graph), $nbpreds_i$ is the total number of predecessors of operation i, and $nbsuccs_i$ is its total number of successors. This formulation of labels does not take into account the operation durations. For complex graphs (presenting several sprays), Equation (6.1) falls in a trap (yielding a poor balancing). The more sprays the graph contains the more its ordering becomes difficult (for a graph having a diagonal adjacency matrix the labelling is very easy, since the graph is 'dense' and uniform). Equation (6.1) was completed to avoid this pitfall as follows:

$$label_i = nbpreds_i - nbsuccs_i + follows_i \tag{6.2}$$

where $follows_i$ is the maximal $label1_i$ value of the direct successors of operation i in the precedence graph. Table 6.1 gives the application of Equations (6.1) and (6.2) to the precedence graph illustrated in Figure 6.3.

For instance, operation 1 has zero predecessors and four successors; thus, $label_1(1) = 0 - 4 = -4$. The direct successors of 1 are $\{3, 4\}$ and their respective labels are $\{-1, 0\}$; thus, the maximal value is ($Follow(1) = 0$). The label of operation 1 according to Equation (6.2) is -4.

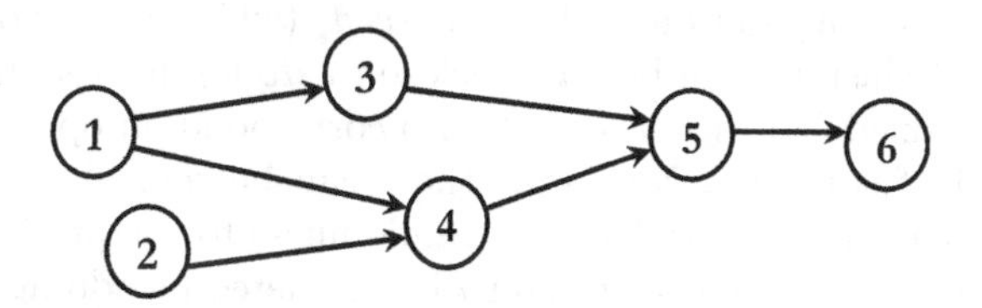

Figure 6.3. Example of precedence graph

This method permits one to order operations by finding the probable first and last operations on the product and permits one to choose the possible seeds of stations (the boundary stones).

Table 6.1. Example of applications of Equations (6.1) and (6.2)

Operation	Duration	nbpreds	nbsuccs	label1	Follows	label
1	4	0	4	−4	0	−4
2	2	0	3	−3	0	−3
3	3	1	2	−1	3	2
4	3	2	2	0	3	3
5	1	4	1	3	5	8
6	5	5	0	5	6[1]	11

Step 3. Boundary stones are chosen using the sequence obtained at the second step. The number of stones is equal to the number of stations. This step allows one to find seeds of piles (stones or stations). In this example, the boundary stones are determined to group operations in three clusters corresponding to the three stations. The first operation in the precedence graph of the product is 1 (it has no predecessor and gets the minimal *label*). The last operation is 6 (having no successor and corresponding to the maximal *label*). According to their labels {−4, −3, 2, 3, 8, 11}, operations are ordered as follows {1, 2, 3, 4, 5, 6} (refer to Table 6.1). The first boundary stone is the label corresponding to the first operation:

$$stone_1 = min(label) \tag{6.3}$$

The boundary stone i is defined as

$$stone_{i+1} = stone_i + gap \tag{6.4}$$

where

$$gap = \frac{\max_i(label_i) - \min_i(label_i)}{N} \tag{6.5}$$

In this example, $gap = 5$ and the boundary stones are {−4, 1, 6}.

Step 4. Once the boundary stones have been fixed, the labels are grouped into as many clusters as stations. The seed $seed_i$ (which corresponds to the first operation) of cluster i will be a *label* close to $stone_i$; for the first station, $seed_1 = label_1$ is fixed. To this $seed_i$ will also correspond an operation which will be the seed of station i. Note that there can be several possible seeds (operations) for each cluster, which adds randomness to the procedure. Once the seeds have been selected, each cluster i is completed by adding $label(s)$ to it in an increasing order, so that

[1] There are no successors for this operation, and it is the last operation in the precedence graph, so the total number of operations is taken as the value of the label.

$$\forall cluster_i, \forall j \in [1, n_{op}], seed_i \leq label_j < seed_{i+1} \tag{6.6}$$

where n_{op} is the total number of operations. The clustering fixes the possible insertion positions (stations) of the remaining unassigned operations. Operations of $cluster_i$ (the one corresponding to the $seed_i$ and the operations in the last cluster excepted) may be assigned to station i or $i+1$. For example, suppose the chosen cluster seeds are $\{-4, 2, 8\}$ and the corresponding label clusters are $\{-4, -3\}$, $\{2, 3\}$ and $\{8, 11\}$. Thus, operation 1 will be assigned to station 1, operation 3 to station 2 and operation 5 to station 3. Among the remaining operations, forming the three clusters $\{2\}$, $\{4\}$, and $\{6\}$ operation 2 may be assigned to stations 1 or 2, operation 4 to station 2 or 3, and operation 6 to station 3, as illustrated in Figure 6.4. The operations already assigned are the station seeds. The arrows starting from the clusters (cl1, cl2, and cl3) point to the station which the remaining operations can be assigned to.

Step 5. Once the clustering has been done the algorithm assigns the remaining operations to stations according to the rules exposed at step 4, taking into account the precedence constraints and the user's preferences (see Section 6.3.3).

Owing to the precedence constraints of the product, some station loads will exceed the desired cycle time (the maximum stations process time). A local improvement phase attempts to equalise those stations again by moving operations along the line or exchanging operations between stations (see Section 6.3.3).

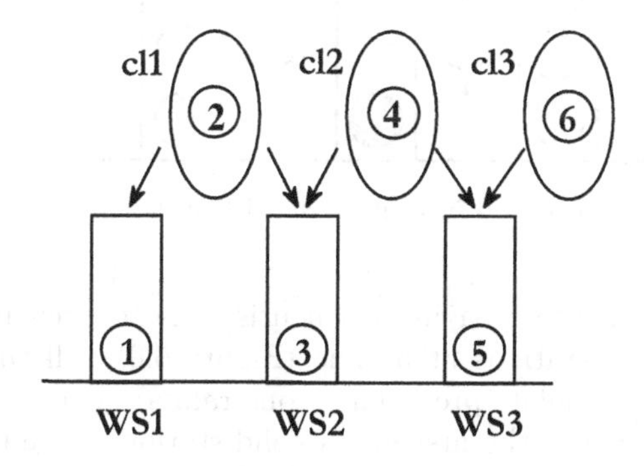

Figure 6.4. Operation clustering and assignment

Steps 4 and 5 are repeated each time to construct a new solution (*e.g.* at the population initialisation) or complete an existing one (*e.g.* after a crossover or during the mutation).

Dealing with Precedence Constraints

At each time, an operation is assigned to a station as follows:

1. Look for the 'boundary stone' that corresponds to the given operation.
2. Have 'X' be the station where to insert the operation. If the stations correspond to the already assigned successors of the given operation that precedes 'X', another station is required in which to insert the operation and to avoid the violation of precedence constraints.

Heuristics

Two heuristics are used to improve the solutions obtained by the boundary stones algorithms: the *simple wheel* and the *multiple wheels.*

Simple Wheel. this heuristic tries to move a set of operations from the first station to the second one. Then, it tries to move a set of operations from the new second station to the third one, and so on until the last station is reached. The move will be accepted automatically if the process time of the destination station added to the processing time of the moved operations does not exceed the cycle time. If it exceeds, the move is accepted with some probabilities. Next, it begins moving operations from the last station toward the last but one, and so on until the first station is reached (see Figure 6.5).

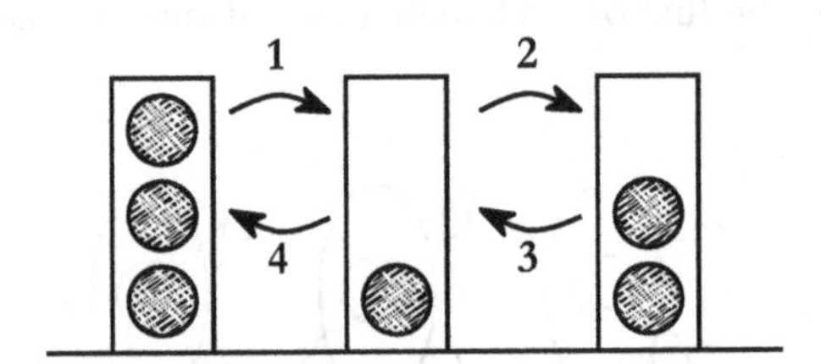

Figure 6.5. Simple wheel heuristic

Multiple Wheel. the idea in this heuristic is to exchange operations between two adjacent stations at a time (Figure 6.6). All the possible exchanges (which do not violate precedence constraints) are executed. The first exchange is made between the first and second station, while the second one is performed between the second and the third station, and so on.

The first heuristic gives the best results, while the second one is used only if the algorithm is stuck in local optima and fails to improve the solution.

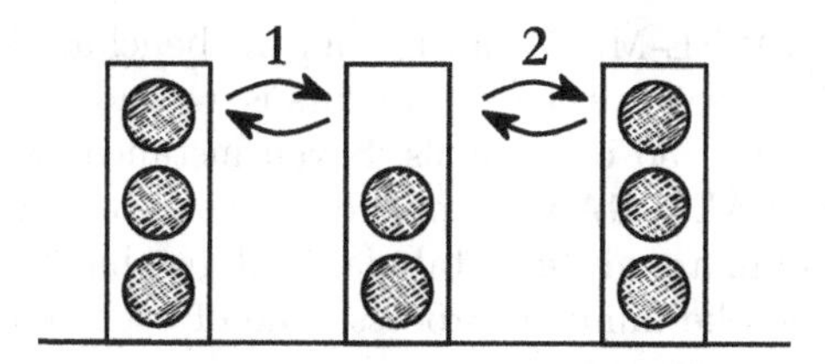

Figure 6.6. Multiple wheel heuristic

Cost Function

The cost function is simply the sum of the differences between the stations, operating times and the desired cycle time. Indeed, a solution where the duration of half the number of stations exceeds the cycle time and the duration of the second half is below the cycle time seems to be a good one, as illustrated in Figure 6.7.

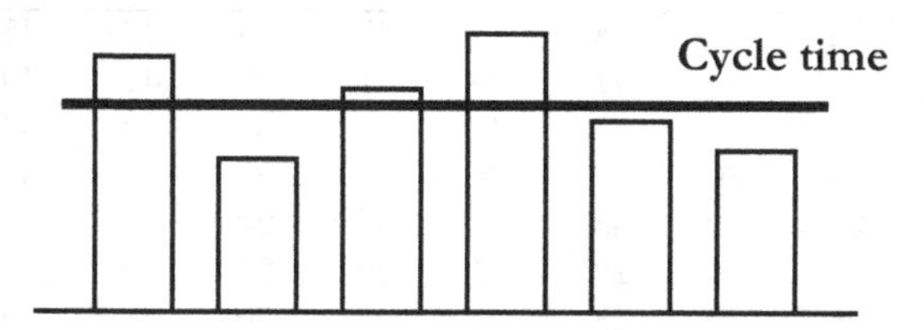

Figure 6.7. Badly balanced line with 'negative' imbalance equal to 'positive' one

The following is proposed to minimise the cost function (balance index):

$$f_{EP} = \sum_{i=1..N} (fill_i - cycletime)^2 \tag{6.7}$$

where N is the number of stations, $fill_i$ is the sum of working times on station i, $time_i$ is the process time of task i, and $cycletime$ is the desired cycle time, defined as

$$cycletime = \frac{\sum_{i=1}^{nop} time_i}{N} \tag{6.8}$$

6.3.4 Experimental Results

In order to assess the merit of the proposed algorithm, many tests have been carried out using standard benchmarks [146]. The proposed metaheuristic was tested on a set of instances (BUXEY, GUNTHER, HAHN, KILBRIDGE,

LUTZ, WARNEKEE, WEE-MAG) taken from the benchmarks[2] proposed by [146]. The EPAL-GGA used in the experiments is a steady-state group-based GA using a population of 36 individuals. Seven instances were used to test the method and the EPAL-GGA was applied five times on each instance. For each instance the results are given in tabular and graphical forms. Table 6.2 gives the maximal and the minimal process time of each solution. Table 6.3 summarises the balancing of the solution.

Table 6.2 summarises the results corresponding to the 'BUXEY' instance. The BUXEY instance is composed of 29 operations and the total process time is 324. CT represents the theoretical cycle time corresponding to a given number of stations. For a given number of stations N, CT is the sum of processing time of all tasks divided by N. Min (Max) represents the minimal (maximal) process time of stations of a given solution.

Table 6.2. BUXEY's minimal/maximal workload of stations

N	CT	Min 1	Max 1	Min 2	Max 2	Min 3	Max 3	Min 4	Max 4	Min 5	Max 5
7	46.2857	45	48	45	48	44	47	44	47	45	48
8	40.5000	40	41	40	41	40	41	40	41	40	41
9	36.0000	32	38	32	38	32	38	32	38	32	38
10	32.4000	27	35	30	34	26	35	27	35	30	35
11	29.4545	20	34	20	33	20	34	20	34	20	33
12	27.0000	23	29	24	29	24	29	24	29	25	28
13	24.9231	20	28	20	28	20	28	20	28	20	28
14	23.1429	20	26	15	26	14	27	20	26	20	26

Table 6.3 gives the balancing which corresponds to the five runs of the best solutions (the balance index was obtained using Equation (6.7) (see Section 6.3.3). The last two columns represent respectively the average run time (on a Pentium II 333 MHz) and the standard deviations of the values. Figure 6.8 represents the balancing versus the number of stations. The balancing obtained is generally less than 1 (1 corresponds to a highly imbalanced AL). Also, the results obtained show that the balancing decreases with the number of stations (non-balanced stations). The bigger the number of stations, the smaller is the cycle time and the smaller is the number of tasks. Thus, the balancing becomes complicated. This shows the complexity of the balancing, as for 14 stations the balancing values are very big, as shown in Figure 6.8. The results obtained using these benchmarks show that the standard deviations are high and the average run time still reasonable, and this proves that the algorithm behaves uniformly.

[2] The benchmark suite is accessible via the Web at http://www.bwl.tu-darmstadt.de/bwl3/forsch/projekte/alb/index.htm

Table 6.3. BUXEY's balancing

N	Balance1	Balance2	Balance3	Balance4	Balance5	Avg Runtime	Std Dev.
7	0.0503	0.0503	0.0589	0.0589	0.0503	29.16	14.774
8	0.0349	0.0349	0.0349	0.0349	0.0349	4.0600	0.171
9	0.1470	0.1470	0.1470	0.1521	0.1470	2.7400	0.475
10	0.2147	0.1525	0.2438	0.2191	0.1757	26.5400	3.087
11	0.3970	0.4168	0.4277	0.4462	0.4140	33.1000	9.507
12	0.2029	0.1737	0.1889	0.1737	0.1283	38.9400	8.911
13	0.3784	0.3654	0.3654	0.3426	0.3698	46.7400	13.073
14	0.3339	0.4567	0.5034	0.3449	0.3225	59.7600	10.598

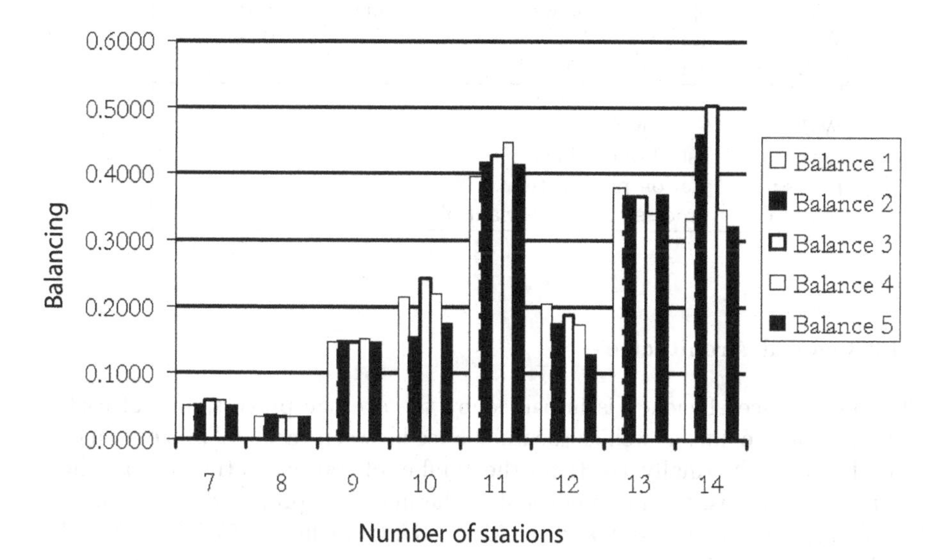

Figure 6.8. The balancing versus the number of stations (BUXEY)

6.4 Extension to Multi-product Assembly Line

6.4.1 Multiple Objective Problem

The multi-product ALB with a fixed number of stations may have two conflicting objectives as follows:

1. equalise the average station load (EPAL);
2. minimise the difference between variants' workload on each station.

The results obtained using one of these objectives are generally different. To illustrate this, suppose we have $M = 4$ operations and $N = 2$ stations and two variants. Suppose the following variant operation's duration ({0, 5},{4, 4},{5, 0},{6, 6}), and the two solutions are given in Table 6.4. The first solution presents a good balancing of the stations on average and a great difference between variants' process times. On the other hand, the second one presents a bad average balancing among the stations but no difference between the two variants' process times. The choice among these two solutions becomes hard and depends on the preferences of the designer.

Table 6.4. Two groupings and their corresponding process times

	op	pt_ws	pt_var1	pt_var2
Ws1	1,2	6.5	4	9
Ws2	3,4	8.5	11	6
	op	pt_ws	pt_var1	pt_var2
Ws1	1,3	5	5	10
Ws2	2,4	10	5	10
WsX :	Station X,			
Op :	list of operations,			
pt_ws :	station process time,			
pt_varY :	process time on variant Y,			

6.4.2 Overall Architecture

The line balance efficiency is impacted by the average process time of each station along the line, as well as the imbalance between variants' process times on each station. Normally, the fewer the number of stations in the line and the less the idle time is, the more efficient is the line. The proposed algorithm is based on the EPAL approach and the MO-GGA (see Chapter 5). This method will be applied each time when constructing or completing the construction of an individual in the GGA. The main steps of the algorithm that serve to generate possible solutions to the problem are summarised below.

1. *Create a population of individuals using EPAL (see Section 6.3).*
2. *Use PROMETHEE II to order individuals in the population.*
3. *Recombine (mate) best individuals (parents) to produce children.*
4. *Mutate children.*
5. *Use PROMETHEE II to order the new population.*
6. *Replace the worst individuals of the population by the new children.*
7. ***If** a satisfactory solution is found stop. **Else** go to 3.*

Input Data

The EPAL for MPAL algorithm needs the following input data, as illustrated on Figure 6.9:

- desired number of stations;
- duration of each operation;
- list of variant products;
- precedence graph of the product family;
- user's preferences.

The first two input data are used to estimate the average cycle time of the assembly line while the third one deals with the mixed-production problem. On the other hand, the last two deal with the user's needs.

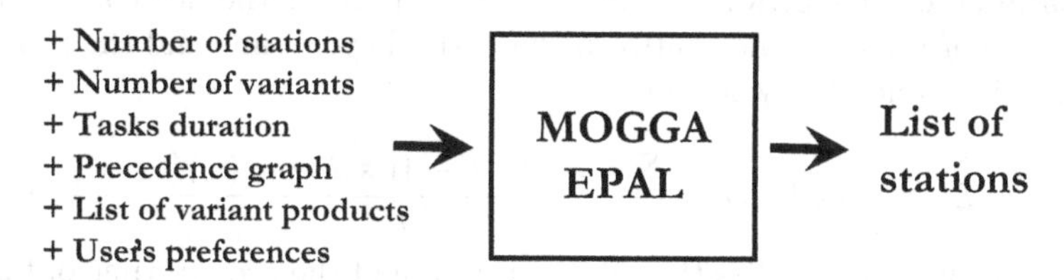

Figure 6.9. Data flow for EPAL

Performance of the Solutions

The two conflicting objectives used to design MPALs are as follows:

Process Time Divergence Among Products Variants. The process time of station w on variant i is given by Equation (6.9). This process time is the sum of the process time of the tasks j of station w on variant i ($Ptime(i,j)$ is the process time of task j on variant i). The number of tasks of each station w is set to M_w.

$$Variant\_Ptime(i,w) = \sum_{j=1..M_w} Ptime(i,j) \tag{6.9}$$

The process time of operation j is the ratio between the sum of process time of task j and the different variants on the number of variants.

$$Operation\_Ptime(j) = \frac{\sum_{i=1..Nb\_Variant}(Ptime(i,j))}{Nb\_Variant} \tag{6.10}$$

The process time of station w is the sum of the process time of the tasks that belong to station w.

$$Ws\_Ptime(w) = \sum_{i=1..N} (Operation\_Ptime(i)) \tag{6.11}$$

Equation (6.12) presents the standard deviation of the process time on station w.

$$Std\_div(w)^2 = \frac{\sum_{i=1..Nb\_Variant}(Ws\_Ptime(w) - Variant\_Ptime(i,w))^2}{Nb\_Variant} \tag{6.12}$$

The line standard deviation of the process time is the ratio between the sum of standard deviations of the different stations and the number of stations.

$$Line\_Std\_Div = \frac{\sum_{i=1..N}(Std\_div(i))}{N} \tag{6.13}$$

Imbalance of the Line. The imbalance of the AL is the ratio of the sum of the square of the difference between the desired cycle time and the process time of stations and the cycle time.

$$Line\_Imbalance = \frac{\sum_{i=1..N}(CT - Ws\_Ptime(i))^2}{CT} \tag{6.14}$$

The ideal cycle time CT is the ratio of the sum of the process time of tasks and the number of tasks N.

$$CT = \frac{\sum_{i=1..N}(Operation\_Ptime(i))}{N} \tag{6.15}$$

where

N	Number of stations.
M_w	Number of operations of station w.
CT	Desired AL cycle time.
$Nb\_Variant$	Number of variants.
$Line\_Std\_Div$	The standard deviation of the AL.
$Line\_Imbalance$	The imbalance among station.
$Std\_Div(i)$	The standard deviation of station i.
$Variant\_Ptime(i,w)$	Process time of station w on variant i.
$Ws\_Ptime(w)$	Station process time.
$Ptime(i,j)$	Process time of operation i on variant j.
$Operation\_Ptime(i)$	Process time for operation i.

The cost function[3] is as follows:

$$CostFunction = PROMETHEE(Line\_Std\_div,\ Line\_Misbalance)$$

[3] Where PROMETHEE(x1, x2, ...) means we use the MCDA method PROMETHEE II, and the objectives are x1, x2, *etc.*

A large imbalance of the workload among different variants has to be avoided. Indeed, even when the load of the stations is balanced on average, the production can easily be faced with work overflow or starvation on individual stations (see Chapter 8).

7

The Resource Planning for Assembly Line

7.1 Introduction

In general, the AL dedicated to *small-sized* products may be *hybrid*, in which the operations can be executed either manually or automatically (see Figure 7.1). Therefore, these lines are called HALs [130]. Given a list of candidate equipment available to complete the operations, the design problem thus becomes to decide which resources to select and which tasks to assign to each resource in order to meet the production requirements at a minimum cost. Many nominations can be found in the literature and the best known are line balancing with processing alternatives, and assembly system design or assembly process planning with resource assignment. The remainder of this chapter is structured as follows. We briefly review the work conducted in Section 7.2. Section 7.3 is devoted to the explanation of the RP, while Section 7.4 is devoted to the input data of the problem. A detailed description of the proposed method is detailed in Section 7.5, and a case study is presented in Section 7.6.

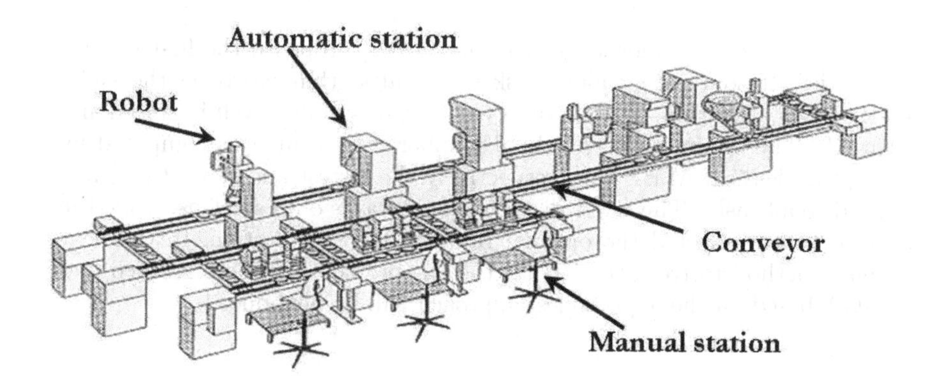

Figure 7.1. HAL illustration

7.2 The State of the Art

Graves and Withney [60] presented a method for the single product equipment selection problem which was solved by a B&B procedure. Graves and Lamar [59] extended the work of Graves and Withney [60] and developed an integer programming cost-based model to the automated system design problem. The approach used a fixed assembly sequence,[1] and the main limitation was that the non-serial line-layouts were permitted [60]. As a result of allowing unrestricted floor layouts for the problem, the solutions found were not necessarily physically realisable. Gustavson [61] developed heuristic methods for solving both the single and multiple product equipment selection problem. Graves and Holmes [58] developed their optimisation method to evaluate the effectiveness of the Gustavson heuristic [61]. The method guarantees the feasibility of the layout by restricting the system to a linear floor layout [62].

Faaland *et al.* [41] also used Gutjahr and Namhauser's heuristic for the ALBP [62]. The authors also proposed two other heuristic adaptations of the shortest path procedure that are capable of solving large problems. Pinto *et al.* [115] discussed processing alternatives in a manual AL (MAL) as an extension of SALB by relaxing the assumption that all stations are identical. The authors proposed a B&B algorithm in which an SALB problem is solved in every node of the tree. This study presented the interesting advantage of working on a set of precedence constraints instead of a fixed sequence. Lee and Johnson [86] proposed an iterative method for MPAL based on integer programming, depth-first B&B and queuing network analysis. The objective was to minimise the cost of work-in-process inventory, machine investment and maintenance and material handling. For more information the reader is referred to [9, 22, 87, 91, 94, 98, 103, 105, 113, 114, 140, 165].

To the best of our knowledge, Falkenauer [42] proposed the first GA for ALB with a 'resource-dependent task times' algorithm based on the GGA and the B&B algorithm. This method was able to supply a well-balanced and cheap AL. The minimal and maximal number of stations was computed by solving the classical ALBP with respectively the fastest and slowest resource assigned to all tasks. The GGA distributed the tasks onto stations, while the B&B algorithm selected the optimal resource for each station. In the next section a method to treat the RP for the AL problem will be presented and MOGGA based on the 'equal piles' approach will be introduced.

---

[1] An assembly sequence is an ordering of m operations to assemble a product; m is the number of components of the product.

7.3 Dealing with Real-world Hybrid Assembly Line Design

A set of possible groups of equipment (feeders, handlers, insertion devices) are called 'functional groups' (FGs) [110]. In the remainder of this chapter a set of FGs are associated to each 'task' and the term 'equipment' means a set of elementary[2] equipment.

The HAL design problem can formulated as follows:

- given a set of tasks and for each task a set of possible resources characterised by their price, reliability, processing time, space, *etc.*
- given the constraints of cycle time and maximum peak time[3] (for variants), possible, precedence among some tasks,

we have to find:

- the resources to be allocated to each task among the possible ones;
- an assignment of tasks to stations along the line, such that:
  - no precedence constraint is violated;
  - the station's workload is as equal as possible to the cycle time of the line;
  - in the case of multiple products the average process time of each station does not exceed the maximum peak time.

The following objectives have to be met (not necessarily all together):

- The total price of resources allocated to tasks is minimal.
- A maximal reliability of the line is attained.
- The surface occupied by the equipment fits the station space.
- The workload of the stations is as balanced as possible.

A set of features is associated with each task 'FG', a cost, a process time, an availability, a space and a list of incompatible tasks.

7.3.1 Cost

The cost of an FG will be given by the sum of the cost of each of its pieces of equipment. The price of a resource is its price over the expected lifespan of the line and it includes: the purchase cost, the exploitation cost added to the maintenance cost, the cost of manpower necessary to use the equipment (including training, *etc.*), and the consumption cost.

[2] Elementary equipment can be any feeding, handling or insertion equipment or any auxiliary operation one can find in ALs (checking, adjusting, cleaning, *etc.*).

[3] This maximum peak time may not be exceeded by any variant process time on a station, while the cycle time must not be overstepped by the average working time on a station.

Only the purchase cost is fixed. The three other costs are variable. In order to evaluate them correctly, it is thus necessary to compute them for a given period of time (a year for instance).

7.3.2 Process Time

There is still a lack of reliable tools for the estimation of process time [18]. Owing to the possible hidden times, it is known that the global duration of an operation is not always the sum of the process times of the equipment in the FG. In order to deal with these kinds of exception, only an interactive constitution of FGs can yield their correct processing time [110]. There are two kinds of hidden times: inside a FG and among several FGs.

Hidden Time Inside a Functional Group

Suppose that we have a group that consists of picking up a part with a manipulator and placing it on a hole where a press will insert it, as illustrated in Figure 7.2. The complete process time of the operation can be sub-divided into:

- visible time – the parts which cannot be executed simultaneously,
- hidden time – the parts which can be executed in parallel.

In this example, the picking up of the part by the manipulator from its storage location can be done simultaneously with the insertion of another part. The insertion phase can begin only when the manipulator can pick up a new part.

The total visible process time should be the sum of the visible time of the two equipments, *i.e.* 6 s + 3 s = 9 s as shown in (Table 7.1) and illustrated in Figure 7.2. This duration of 9 s being greater than 8 s is necessary for the manipulator to accomplish a complete operating cycle. The actual operating time of the group to consider when balancing the line will thus be 9 s.

Table 7.1. Subdivision of the operating times

	Press	Manipulator
Visible time (s)	3	6
Hidden time (s)	0	2
Total Time (s)	3	8

The operating time of a given group will thus be given by the sum of the visible times of each equipment or by the largest total operating time of a piece of equipment in the group if it is larger than that sum [110].

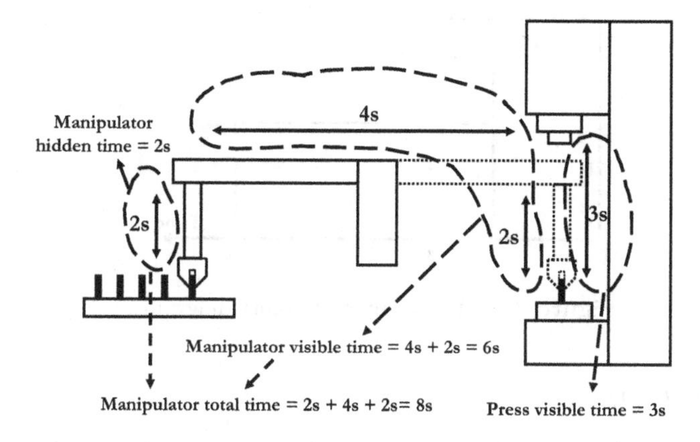

Figure 7.2. Illustration of the process times decomposition [110]

Hidden Times Between Functional Groups

Figure 7.3 shows two different FGs (FG_1 and FG_2) coupled on the same station which could simultaneously execute their corresponding operations. The process time of the station will therefore not be equal to the sum of the process time of the FGs, but rather to the larger process time of the two FGs. The designer usually uses Gantt diagram, as illustrated in Figure 7.4 to estimate the resulting process time. Indeed, in this example, FG_2 starts 2 s after FG_1 (*i.e.* 1 s before the end of the process cycle of FG_1). Therefore, it seems obvious that the exact process time of the two FGs cannot be automatically pre-determined using simple rules is as the case for the first kind of hidden times. Figure 7.4 shows an example in which the exception list is $\{(FG_1, FG_2), 6\}$. Thus, if a station contains FG_1 and FG_2 the process time is set to 6 and not the sum of their process time which is 8 $(2 + 6)$.

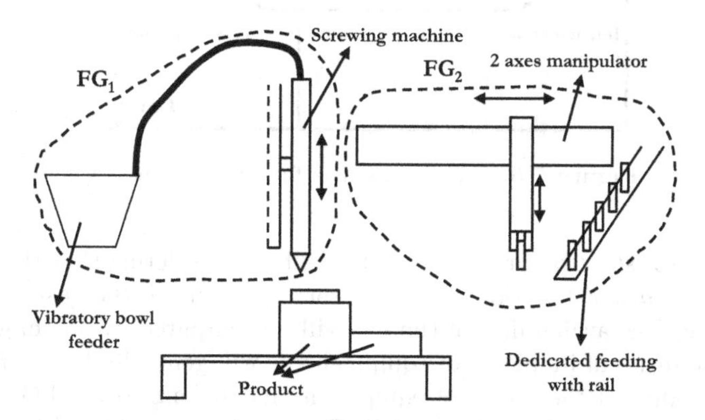

Figure 7.3. The coupling of FGs on the same station [110]

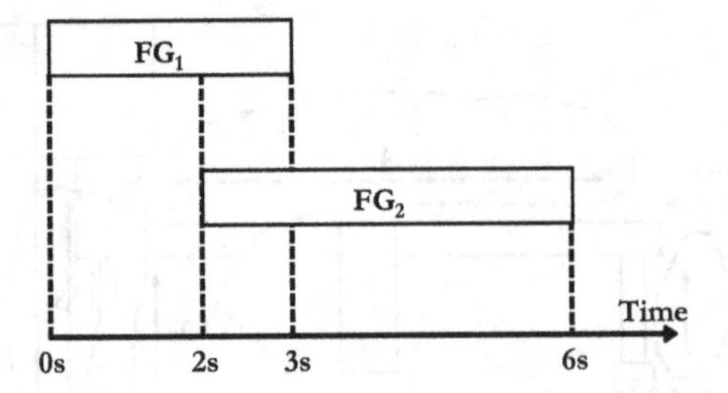

Figure 7.4. An example of Gantt diagram

7.3.3 Availability

The availability of a piece of equipment is defined as the proportion of total time that is available for use. Therefore, the availability of a repairable piece of equipment is a function of its failure rate and of its repair or replacement rate [104]. Figure 7.5 shows the evolution of the failure rate during the lifetime of the equipment, which is defined by the so-called *bath-tub curve* [120]. It is clear that the failure rate depends on the lifetime t. Three periods can be distinguished from this curve:

1. the *infant mortality period*, in which the failure rate drops;
2. the *constant failure rate period*, in which accidental failures occur;
3. the *wear-out period*, in which the failure rate rises due to technical age.

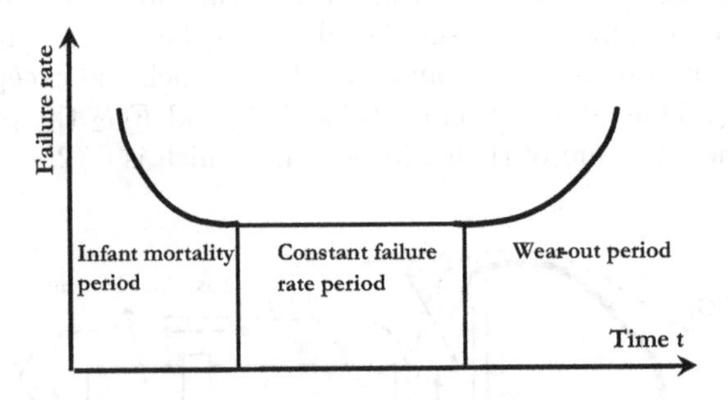

Figure 7.5. Failure rate as a function of time

Assume a *constant* failure rate is equivalent to considering that the equipment is in its *normal* functioning period corresponding to the vast majority of its lifetime. The availability of the FG will be computed by a combination of the availabilities of its pieces of equipment. It will generally be the *product* of the availability of the pieces of equipment that belong to the FG as they have a *serial* configuration (Figure 7.6). The availability of the FG is equal to $Av_1 \times Av_2 \times Av_3$. The dependence between these pieces of equipment was neglected. This assumption can correspond to the large majority of cases, but we are aware of its limitations. Thus, further research on the way to model such systems will be fully encouraged.

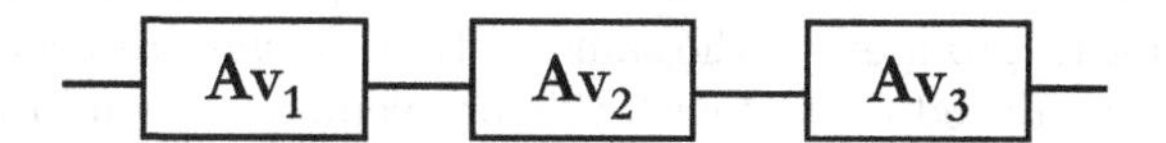

Figure 7.6. Representation of an FG as an serial system

7.3.4 Station Space

Station space is proportional to the space for storage of parts used in the station as well as the space occupied by the equipment. The storage space (Storage_Space) is the space needed to store the parts before being assembled. A constant C_i is assigned to each equipment and it is proportional to the space that it occupies. Figure 7.7 shows a simple representation of the space occupied by a set of equipment. Suppose that the station is composed of k equipment units, then the station space (Station_Space) is given by

$$Station\_Space = (\sum_{i=1}^{k} C_i) + Storage\_Space$$

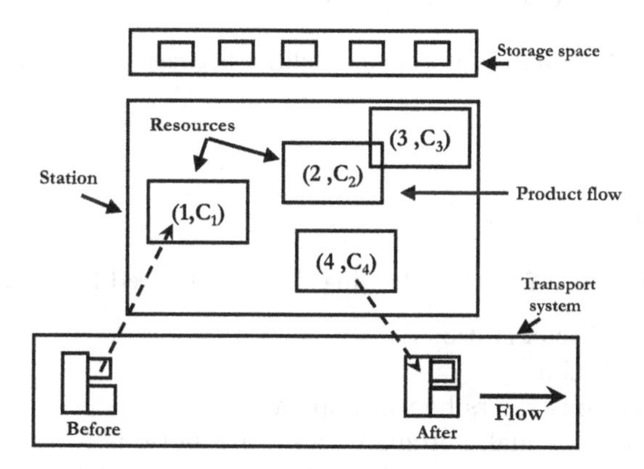

Figure 7.7. Station space

In order to really estimate the space occupied by a set of equipment, more information is needed on the shape of the equipment as well as the shape of the station. In this case, the '2D-bin packing' problem arises, and this leads to the physical layout problem which is one step more in the design of AL, namely layout.

7.3.5 Incompatibilities Among Several Types of Equipment

Another element that has to be considered is the possible incompatibility between certain kinds of equipment. Indeed, it can happen that two categories of equipment cannot be grouped on the same station. A *matrix* of incompatibilities called I is introduced to deal with this problem, as shown in Figure 7.8. The size of the symmetric matrix I is equal to $N \times N$, where N is the number of equipments and each I_{ij} element of that matrix will have the following value:

- $I_{ij} = 0$ if equipment i is compatible with equipment j;
- $I_{ij} = 1$ if equipment i is not compatible with equipment j.

	Eq_1	Eq_2	Eq_3	...	Eq_N
Eq_1	1	0	0	...	1
Eq_2	0	1	1	...	1
Eq_3	0	1	1	...	1
⋮	⋮	⋮	⋮	⋱	⋮
Eq_N	1	1	0	...	1

Figure 7.8. Matrix of incompatibilities between equipments

7.4 Input Data

As illustrated in Figure 7.9, the following data are nedeed to design an HAL:

- the desired number of stations;
- the desired cycle time;
- the precedence constraints between operations;
- operation mode (manual, automated and/or robotic);
- the list of exceptions to deal with hidden times;
- an equipment database which yields the features of the different resources (cost, reliability, process time, equipment space);
- the list of its incompatible equipment.

The preparation of data, and especially the 'equipment selection' step, yields to a set of equipment that can perform the given set of operations. Economic criteria are used in the evaluation of the equipment selection process [110].

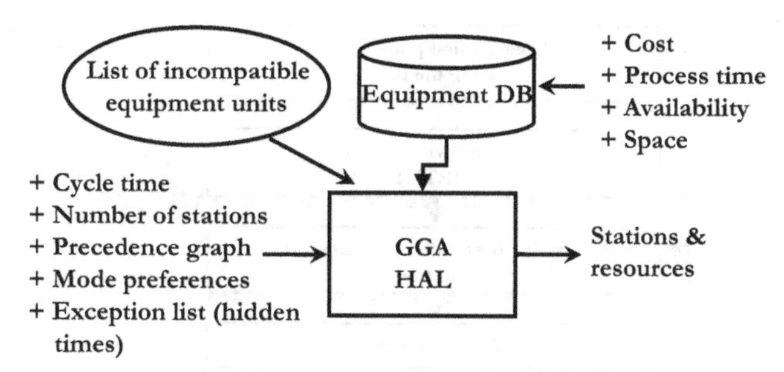

Figure 7.9. Data flow for an HAL

7.5 Overall Method

The following algorithm is proposed to generate possible solutions of the problem [122]. The ICA is used each time when constructing a solution for a problem as follows:

1. Assign tasks to the stations (using the process time corresponding to the fastest equipment) according to an equal piles strategy.
2. Generate all possible resource combinations using a branch & cut (B&C) algorithm.
3. Select the best equipment combination using PROMETHEE II.

Since the objective is to deal with many conflicting objectives, the MOGGA approach is adopted (Figure 7.10), as explained in Chapter 5.

7.5.1 Distributing Tasks Among Stations

In this section we recall the strategy used to group tasks and introduce the operating mode as a new constraint.

Equal Piles Algorithm

In order to assign operations to stations, the EPAL is used. At the operations-to-stations stage, a minimum cycle time *min_ct* is used. This *min_ct* is the ratio between the sum of minimum process times of the operations and the desired number of stations (see Section 6.3).

Mode Preferences

The associative constraints can be defined as '*The manual tasks have to be grouped together, and the robotic or automated tasks have to be grouped together*', while the dissociative preference constraints can be defined as '*The manual tasks cannot be grouped with robotic or automated tasks*'. Those constraints are said to be hard because they cannot be violated.

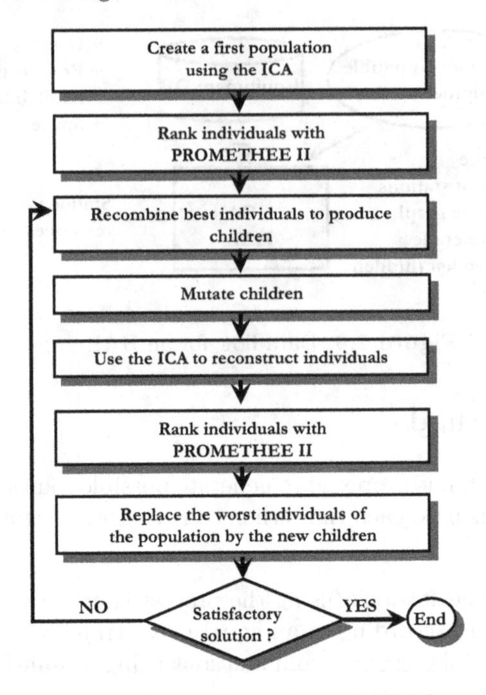

Figure 7.10. Steps of the MOGGA

7.5.2 Selecting Equipment

Owing to the cycle time constraint, the resources must be fast enough to perform all the tasks on time. This allows one to define a B&C procedure that efficiently explores the 'huge' search space. Consequently, even for large problem instances, the B&C typically handles only a small number of tasks even with the fastest resources. Thus, the size of the B&C problem stays reasonably small and leads to an acceptable speed of the method. The MCDA method of PROMETHEE II is used to deal with the different objectives addressed by the designer. The main features of the algorithm can be summarised in the following steps:

develop the first node (task) level i=0;
verify the validity of the offspring nodes;
repeat
  if there is no valid nodes then stop;
  generate all the offspring(s) of all valid nodes : create level i+1;
  verify the validity of the offspring's nodes of level i+1;
until *the last level is attained;*
use PROMETHEE II to choose the best solution among the valid ones.

Branch and Cut Algorithm

The B&C method consists of two fundamental procedures: branching and cutting. Branching is used to divide a large problem into two or more sub-problems that are usually mutually exclusive. A branch is associated with each sub-problem. These can be partitioned in a similar way, yielding new branches. The partitioning process stops if it represents only one solution, or if it can be shown that the node does not contain an optimal solution. The B&C algorithm is used to assign equipment to operations. Each node corresponds to a piece of equipment and each level to one task. On the graph presented in Figure 7.11, each couple (a, b) corresponds to an equipment and the sum of process times of this branch at a given level. For example, the couple $(5, 11)$ means that, by selecting equipment 5 to realise operation 2, the total process time of the station is 11. At each level, all possible equipment of the given operation is generated, but only the valid branches respecting the constraints of the problem are developed further. For instance, selection of equipment 2 for operation 1 and equipment 6 for operation 2 yields a process time of 16. Since the cycle time is set to 15, this branch will not be developed further. Once all the levels (valid branches) have been developed, only valid solutions are kept. Valid solutions are the solutions that verify the following constraints:

- the sum of process times of operations for the selected equipment must not exceed the cycle time;
- the list of equipment used at each station must not be incompatible.

Suppose we have n tasks and each task i has b_i possible pieces of equipment. The number of levels equals n, which is the number of tasks, while the number of nodes is given by

$$NbNods = \sum_{j=1}^{n} (\prod_{i=0}^{j} b_i) \times b_j \qquad with \quad b_0 = 1;$$

The number of end-leafs is equal to

$$NbEndLeaf = \prod_{i=0}^{n} b_i$$

For instance, in the case $n = 3$ and $b_i = 2$ for $i = \{0, 1, 2\}$, the number of nodes is 14 and the number of end-leafs is 8. The *cutting* mechanism is used to save time while exploring the tree. The following decisions have a great influence on the run time of the cutting algorithm: How the tasks are assigned to the different levels. How the equipment is ordered for each task. Since there is a set of tests of validity of the solution to be done, deciding which one to begin with.

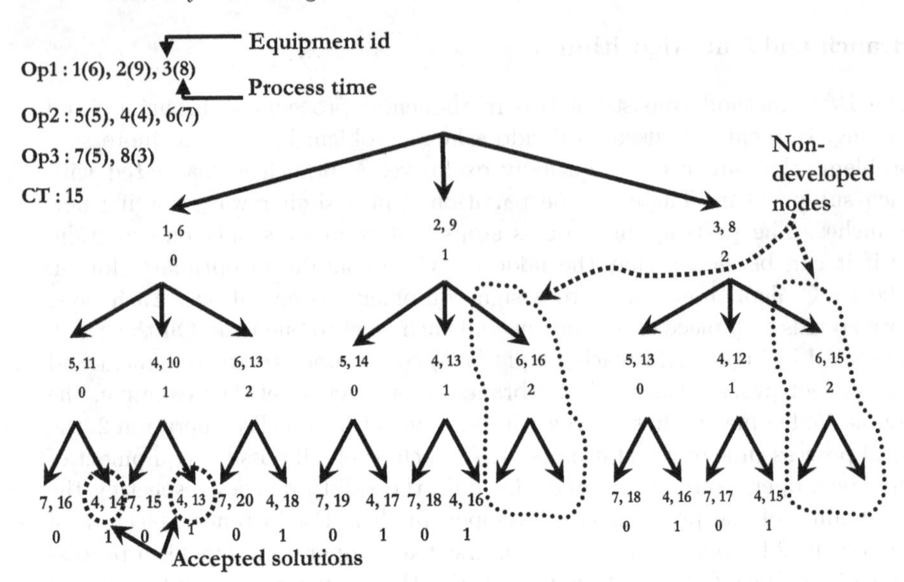

Figure 7.11. Tree generated by the B&C algorithm

How and Where Does the User Intervene?

The criteria adopted during the selection of equipment is as follows:

- Process time for each station should not require more than a cycle time to perform all the tasks.
- Minimise cost of the resources allocated to the stations.
- Maximise reliability on each station.
- Reduce space proportional to the space occupied by the station.

Thus, once all the end-leafs of the tree have been found, the time comes to choose the best solution among the valid ones. Since each solution is characterised by its cost, process time, and reliability, classical *pairwise* methods to compare solutions cannot be used. The different solutions found by the B&C algorithm serve as input data for the PROMETHEE II method to choose the best equipment taking into account the different criteria. Afterwards, resources are assigned to each task of a given station. A non-valid solution can be:

- a solution in which the sum of process times of the fastest equipment exceeds the cycle time;
- a solution which is composed by only incompatible equipments;
- a solution in which the desired cycle time is incompatible with the fixed number of stations;
- a solution in which too many tasks are grouped on a given station.

If there is no possible solution among the nodes developed, the solution which corresponds to the fastest equipment will be selected.

7.5.3 Heuristics

Two new heuristics are introduced to deal with the hard constraint of operating mode of the tasks. These heuristics are (1) the merge and split, and (2) the pressure difference heuristic.

Merge and Split. Figure 7.12(a) presents the situation when dealing with the operating mode of HAL. Suppose there are two non-filled adjacent manual stations and an over-filled automated one. In order to find a good balancing, one way is to merge the two manual stations and to split the automated one. Figure 7.12(b) presents the solution obtained after the merge and split procedure. The result is one manual and two automated stations. From the balancing point of view, the second solution is better than the first one. Note that the sum of the process times of the two new automated stations (70% and 90%) is not necessarily the process time of original (130%).

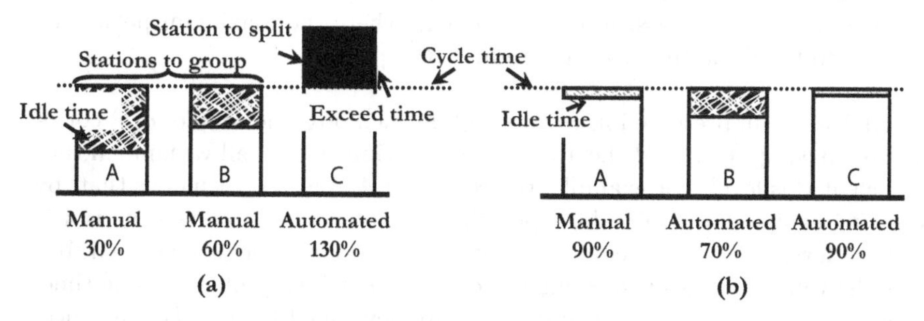

Figure 7.12. Solution before (a) and after (b) the merge and split heuristic

Pressure Difference. The goal is to start the search at a station that exceeds the cycle time (the high pressure) as well the station that is less filled (less pressure). The goal is to move the excesive process time of station C in Figure 7.13(a) to fill the gap (idle time) existing in station A. The operating mode and the precedence constraints of the tasks have to be verified. In this case, a task i to be moved from station C must have all its predecessors in station A (or before). If the move from A to C, all the successors of task i would have to be in C or later. Figure 7.13(b) presents the solution obtained after executing the procedure. The kind and the number of stations obtained is the same as before the application of the heuristic. The simple wheel and multiple wheel heuristics cannot improve such a solution, since the two manual stations are separated by an automated one. It is clear that the second solution is better balanced than the first one.

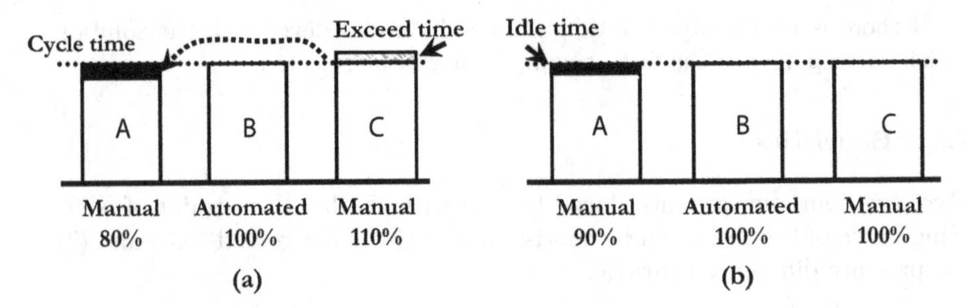

Figure 7.13. Solution before (a) and after (b) the pressure difference heuristic

7.5.4 Dealing with a Multi-product Assembly Line

In this section, the method used to deal with multi-product resource planning is developed. The main goal is to find the cheapest assembly system. A *maximum peak time* parameter fixed by the designer is introduced to allow some variants' process times to exceed the desired cycle time. This maximum peak time may not be exceeded by any variant process time of any station, while the desired cycle time must not be overstepped by the average working time on any station. The classical case is the one where the maximum peak time is equal to the cycle time (single product RP) [132].

On the graph presented in Figure 7.14, in each box, the couple (a, b) corresponds to equipment, and the average process time among all variants in that branch at this level. For example, the couple $(2, 4.5)$ at level 1 means that, by selecting equipment 2 to realise operation 1, the average process time is 4.5. At each level, all possible equipment of the given operation is generated, but only the valid branches respecting the constraints of the problem (cycle time, maximum peak time, compatibility, *etc.*) are developed further. For instance, selecting equipment 1 for operation 1 and equipment 6 for operation 2 yields an average process time of 9.5 and a process time of 13 on variant 1 and a process time of 6 on variant 2. Even if the average process time (9.5) is less than the cycle time (10), since the process time of variant (13) exceeds the maximum peak time, this branch is not valid. Using equipment 1 for operation 1 and equipment 4 or 5 for operation 2 leads to a valid solution. In order to select the best set of resources, once again the PROMETHEE II method will be used. Valid solutions are the solutions that verify the following constraints:

- the average process time of operations of the selected equipment must not exceed (on average) the cycle time;
- the process time of operations of any of the *variants* for the selected equipment must not exceed the 'maximum peak time';
- the equipment used at each station must not be incompatible.

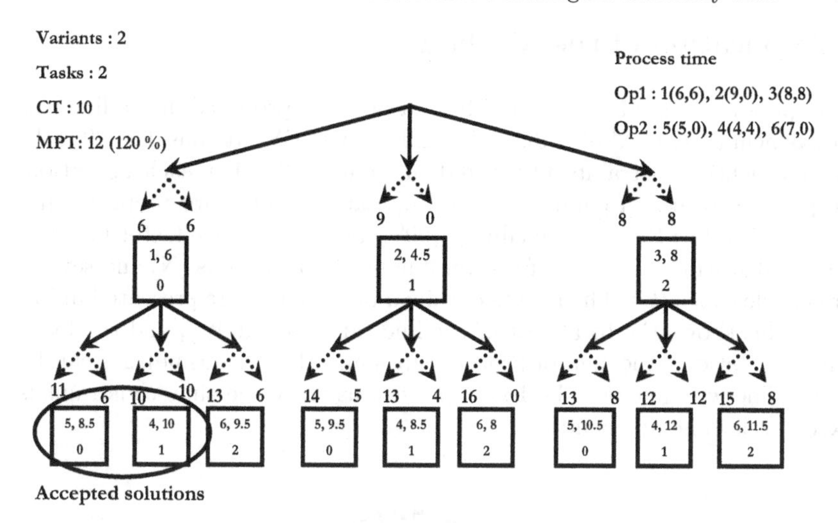

Figure 7.14. Multi-products tree generated by the B&C algorithm

7.5.5 Complying with Hard Constraints

The GGA developed deals only with valid solutions that were tested at each level (*i.e.* constraints and preferences introduced in Section 7.3). The method verifies incompatibilities among equipment, and updates the process time of the stations using the hidden times, *etc.* The main features of the method can be summarised in the following steps.

repeat *for each level*
- *evaluate the process time of each variant, the average process time of station,*
- *evaluate cost, reliability, stations space;*

    repeat *for all nodes*
    - *cut the branch if the used equipment are incompatible;*
    - *update the process time using the masked time;*
        + *if the average process time exceeds the cycle time, cut the branch;*
        + *if the process time of a given variant exceeds the maximum process time, cut the branch;*

    until *the last node is attained;*

until *the last level is attained.*

7.6 Application of the Method

The proposed case study is adapted from a problem proposed in the line balancing benchmark suite[4] of [146] and is called 'BUXEY'. It considers 29 tasks with precedence constraints illustrated in Figure 7.15. For each operation, three possible pieces of equipment (and operating times) are proposed and this equipment has the same reliability (99%) and same space factor (1). The proposed algorithm was tested for several numbers of stations (N) and several desired cycle times (C). The results obtained of the GGA are presented in Table 7.2. This shows the total cost of the line (arbitrary units), and the loads of stations (ratio of the sum of process times and the cycle time). As can be seen, the line will generally be less expensive as the cycle time constraint is relaxed (cycle time augments).

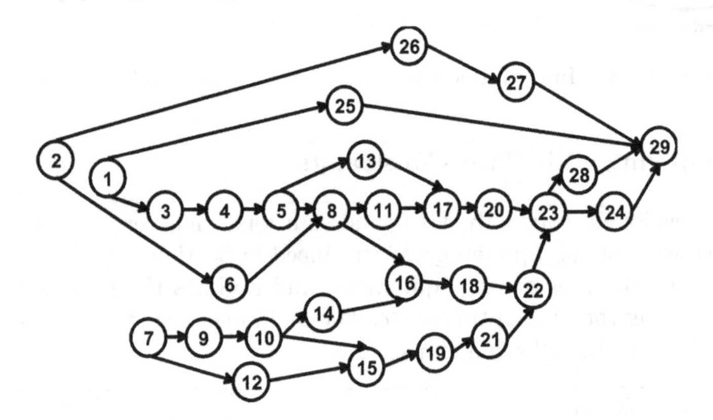

Figure 7.15. Precedence graph of the problem

Table 7.2. Results of the HAL balancing algorithm

N, C	Cost	Station loads
6, 44	3340	1.00, 1.00, 1.05, 1.07, 1.05, 1.00
6, 45	3340	1.00, 1.00, 1.00, 1.00, 1.00, 1.00
6, 46	3280	1.00, 1.00, 1.00, 1.00, 1.00, 1.00
7, 38	3230	1.00, 1.05, 1.03, 0.97, 1.05, 1.08, 1.00
7, 39	3240	1.00, 1.03, 1.03, 1.00, 1.03, 1.00, 1.00
7, 40	3270	1.00, 1.00, 1.00, 1.00, 1.00, 1.00, 1.00
8, 34	3280	1.00, 1.00, 1.03, 1.00, 1.03, 1.00, 1.00, 1.00
8, 35	3240	1.00, 1.00, 1.00, 1.00, 1.00, 1.00, 1.00, 1.00
8, 36	3030	1.00, 1.00, 1.00, 1.00, 1.00, 1.00, 1.00, 1.00

[4] The benchmark suite can be accessed via the Web at
http://www.bwl.tu-darmstadt.de/bwl3/forsch/projekte/alb/index.htm

8

Balance for Operation

8.1 Introduction

Producing multiple products in a single line yields at least two major problems that have to be solved: getting the right input to the right place at the right time, and determining the best production run to match supply with demand. Several technical problems are associated with the design and operation of an MPAL; the most important ones consist of generic product modelling, operating modes and assembly techniques, line layout and model launching (ordering variants) [27]. Model launching (ML) is concerned with scheduling the different models to be produced during a given work shift. In this chapter, a concurrent strategy for product family and assembly system development will be presented. Section 8.2 presents general features of the mixed-production AL design problem. Related studies concerning design of a multi-product AL are presented in Section 8.3. The essentials of ordering GAs (OGAs) are given in Section 8.4. The general architecture of the balance for operation concept is introduced in Section 8.5.

8.2 Multi-product Assembly Line

Local buffers are introduced in order to maintain the pace of the line as even as possible. Each station has a local buffer[1] (upstream). Stations process one unit at time and are linked to buffers by conveyors. When a station completes a process, a product is moved to a downstream buffer if possible (sufficient place); in the other case it remains at the station until it can move. Once a station is free, it takes a product from the upstream buffer (if it is possible)

[1] Buffers allow stations to operate independently, cushioning against machine failure, worker or part shortage, and production rate difference. Large buffers increase throughput time, space requirements, material handling costs, *etc.* Thus, the buffer size must be reduced as much as possible.

and processes a new job. The ordering variants problem occurs when a number of products (variants) are produced on an AL at the same time. Each job[2] must be processed by each station exactly once. Furthermore, all jobs have the same routing (*i.e.* they must visit the stations in the same order). Without loss of generality, we can number the stations so that station 1 is first, station 2 is second, and so forth. A job cannot be processed on the second station until it has completely processed on the first station. The objective is to determine the sequence of variants which maximises the utilisation of the assembly stations.

Chevalier et al. [25] pointed out that there is a strong link between how a line is designed and how it can be operated. The authors investigated how models from the mechanical engineering literature can be combined with models originating from the management science literature. In its most general form, the ML problem is defined by the following: products to be produced, tasks that must be executed on the different products, stations on which a set of tasks have to be performed, constraints which must be satisfied, and the measures to judge the schedule performances. Off-line scheduling (called static scheduling) refers to the formation of a complete schedule before launching production. On-line scheduling (called dynamic scheduling) is performed incrementally while the facility is operating. As each decision problem arises, the results are immediately evaluated and a choice is made [137].

8.3 The State of the Art

It is usual to find MPALs in industry in which the sequencing of models is done without applying any of the heuristics introduced below. Instead, simple rules-of-thumb and experience factors influence selection of the model sequence. For example, if the total production is 100 units per day, and includes 20 units of variant V1, 10 units of variant V2, and so forth, then every fifth unit launched will be a variant V1, every tenth unit will be a variant V2, and so on. The first research that addressed the MPAL sequencing problem was apparently presented in [172]. Most scheduling problems belong to the class of NP-hard problems. Most research has been focused on either simplifying the scheduling problem or devising efficient heuristics for finding acceptable (not necessarily optimal) solutions.

8.3.1 Classical Methods

Monden [100] introduced a goal chasing (GC) method which is based on the part usage goal. The objective was to keep a constant speed in the consumption of each part on the MPAL. Berger *et al.* [15] developed a B&B algorithm

[2] A job is defined as an activity that transforms inputs (a set of requirements) to outputs (products to meet those requirements).

for solving a *tree*[3] ALB problem (TALB) which is a special case of the MPALB problem. The precedence graph is not necessarily a *forest*,[4] but may be any directed acyclic graph. The aim is to minimise the number of stations necessary to manufacture all products. Bard *et al.* [10] formulated the sequencing problem as an integer programming method in order to establish a common mathematical framework that might be applicable to various MPAL configurations. For more information the reader is referred to [3, 82, 96, 100, 144].

8.4 Heuristics

Driscoll and Abdel-Shaffi [40] presented an integrated line balancing and simulation based evaluation technique to address the ALB problem taking into account stochastic task durations, mixed-model processing, task times greater than cycle time, and zoning requirements. The technique first performs a line balance using the RPW technique [65]; it then performs simulations to assess the performance of the layout. Wang and Wilson [170] compared several AL designs in terms of station idle time, incomplete units, and production rate. They proposed a sequencing heuristic, while a simulation was used to evaluate the performance of the different solutions. Fernandez and Groover [47] proposed a mixed-ML algorithm. The objective is to minimise the sum of squares of the deviations from a perfect model sequence in which idle time and work congestion are both zero. The principal problems in the design and operation of ALs are discussed, in addition to the methods for their solutions. For more information, the reader is referred to [5, 11, 21, 26, 64, 71, 93, 99, 137, 146, 162, 168].

8.5 Ordering Genetic Algorithm

The ordering variants problem occurs when a number of products (variants) are produced at the same time. In the following, a GA for the multi-product scheduling problem is presented.

8.5.1 Algorithm

The OGA is used to schedule variants on an MPAL; different heuristics are also used as initialisation techniques.

---

[3] Trees are the nontrivial recursively defined objects: a tree is either empty or a root node connected to a sequence (or a multi-set) of trees.

[4] A forest is a number of disjoint trees.

Encoding Scheme

Because the ordering variants problem is essentially a permutation problem, the permutation of jobs is used as a representation scheme of solutions, which is the natural representation[5] for sequencing problems. For example, let the genes of the given chromosome be $C = [1\,2\,1\,3\,1\,2]$. This means that the job sequence is $v1$, $v2$, $v1$, $v3$, $v1$, $v2$ (see Figure 8.1).

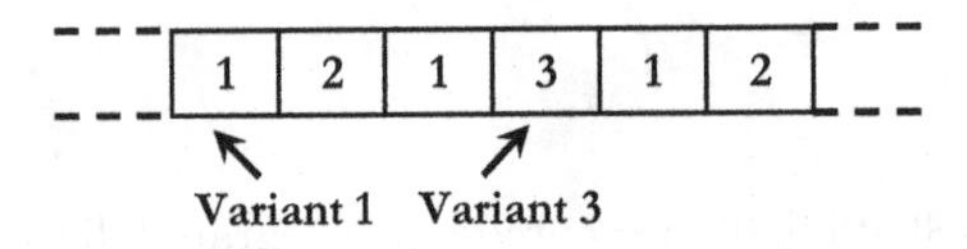

Figure 8.1. An ordering of variants and the corresponding OGA chromosome

Crossover

Various techniques are known from the literature, like partially mapped crossover (PMX) [57], order crossover (OX) [32] and position based crossover (PBX) [54]. In this book, the PMX and the modified PBX are used, as they suit the constraints of the problem.

Mutation

According to the nature of the ordering problem, one or more of the following operators can be applied as follows:

- Shift the place of a randomly selected jobs.
- Invert a place of two selected jobs.
- Shift a selected job from position 1 to position 2.
- Select a job and find a place to insert it so as to minimise the makespan.

Inversion

The inversion serves to shorten promising schemata made of co-adapted genes. A segment on the chromosome is selected at random and the order of genes in that segment is inverted [69].

---

[5] The choice of representation controls the size of the search space. If one chooses a very general representation, more types of problem may be solved at the expense of searching a larger space. Conversely, one may choose a very specific representation that significantly reduces the size of the search, but much will work on only a single problem instance [30].

Evaluation

Since the aim is to minimise the total production time, a simple way to determine the fitness for each chromosome is to use the makespan:

$$Eval(chromosome) = makespan$$

A fast *simulation* procedure (algorithm) is used to estimate the total assembly time (makespan) of a given mix (permutation of variants). Variants are introduced in the order given by their corresponding chromosome (first in, first out). Set-up time is taken into account while estimating production time of each variant (*i.e.* if *variant 1* is followed by *variant 2*, set-up time is added to operating time of *variant 2*). The total production time is reached when the last operator (last station) finishes its job on the last variant of the mix. Idle time is the time lost by operators (machines or robots) waiting for jobs, and it is calculated as follows:

$$IdleTime_{i,j} = Begin_{i,j} - End_{i,j-1}$$

where $IdleTime_{i,j}$ is time lost by operators waiting for job j at station i, $Begin_{i,j}$ is the beginning production time of variant j on station i, and $End_{i,j}$ is the end production time of variant j on station i.

The makespan is given by

$$Makespan = End_{lw,lj} - Begin_{1,1}$$

where $Makespan$ is total production time of the mix, $End_{lw,lj}$ is the end production time of the last station on the last job, lw is last station of the assembly line, and lj is last job of the mix of products.

8.5.2 Heuristics

Several heuristics are used to construct valid solutions. They are used each time when constructing new solutions or to improve the quality of existing ones.

Random. Jobs are inserted randomly in chromosomes. This permits one to avoid local optima.

Batch by batch. If the total production is 100 units per day, which includes 20 units of variant V1, 10 units of variant V2, and so forth, the 20 units of type V1 will be launched, then the 10 units of type V2, and so on.

Mix percent. If the total production is 100 units per day, which includes 20 units of variant V1, 10 units of variant V2, and so forth, every fifth unit will be launched a job of type V1, every tenth unit will be a job of type V2, and so on.

Slope order index. The idea is to give higher priority to jobs with processing times that tend to increase from station to station, while jobs with processing times that tend to decrease from station to station will receive lower priority [54]. The slope index s_i for job i is calculated as

$$s_i = \sum_{j=1}^{m} (2j - m - 1)t_{ij}$$

where t_{ij} is the process time of station j on task i, m the number of stations and n number of jobs. A permutation schedule is contructed by sequencing the jobs in a non-increasing order of s_i, such as $s_{i1} \geq s_{i2} \geq ... \geq s_{in}$.

Gupta's heuristic. Is similar to the slope order index heuristic, except that it takes into account some interesting facts about optimality of Johnson's [75] rule for the three-station problem [54]. The slope index s_i for job i is calculated as follows:

$$s_i = \frac{e_i}{\min_{1 \leq k \leq m-1}\{t_{i,k} + t_{i,k+1}\}} \quad where\ e_i = \begin{cases} 1 & if \quad t_{i1} < t_{im} \\ -1 & if \quad t_{i1} > t_{im} \end{cases}$$

Thereafter, the jobs are sequenced according to the slope index s_i.

The multi-product AL has the following features:

- Each task is assigned to one station.
- Production is composed of a mix of variants. The quantity of each variant is known at the beginning (*i.e.* we have to produce 20 products of variant type 1 and 50 products of variant 2, and so forth).
- Process time of variants can exceed cycle time, but the average on all variants cannot (maximum peak time constraint).
- The job sequence at the entry of stations is the same (first in, first out). When tasks are performed on manual ALs, times to perform each task vary from cycle to cycle. Experience indicates that task times can be approximated by a normal distribution when operators work under paced conditions.
- A fast simulation algorithm is used to evaluate a makespan of solutions.
- The size of buffers is finite.

The schedule is represented using a 'Gantt chart' (Figure 8.2). The x axis corresponds to the time and each horizontal bar corresponds to a station. When a job is processed on a station, a rectangle is placed on the horizontal bar, which begins at the start time of the job and ends at its completion time.

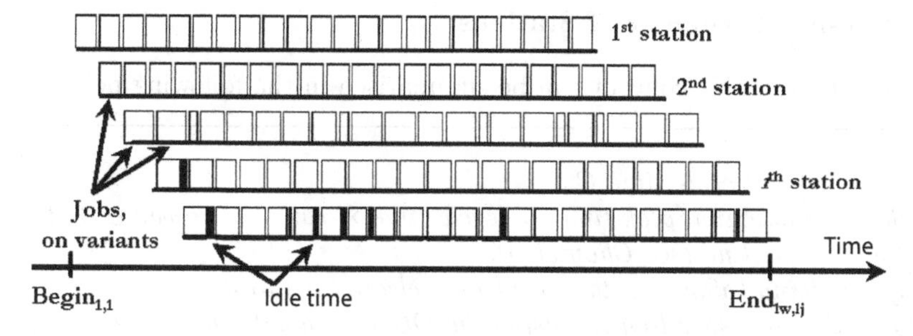

Figure 8.2. Gantt chart for schedule

8.6 Balance for Operation Concept

The BFO concept is introduced to tackle the operation phase problems (ordering) at the design stage. The balancing of the line is realised using the *desired cycle time* C (production forecast). The *effective cycle time* C_E of an MPAL is defined as

$$C_E = average(\frac{makespan}{size\,of\,mix})$$

where the size of the mix is the total number of variants produced in a given period. This balancing phase is iterative, since it is difficult to find a good maximum peak time for a given AL due to the problem's constraints. The ordering algorithm aims to minimise the makespan, and consequently the effective cycle time C_E. The balance for operation is an iterative and interactive procedure used to balance the MPAL, taking the scheduling into account, as illustrated in Figure 8.3. The whole procedure of the BFO for the two approaches is described in the next section.

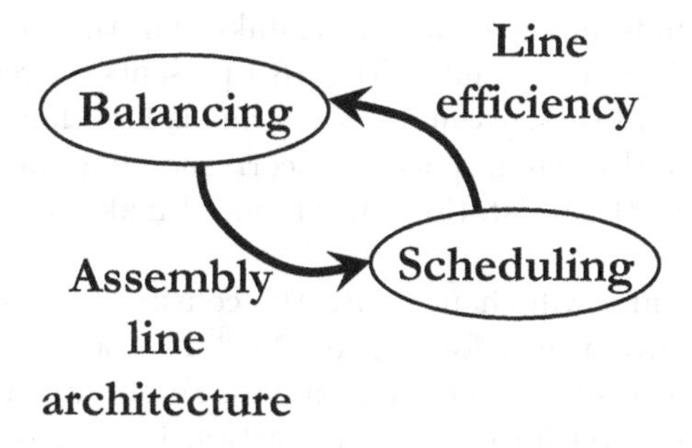

Figure 8.3. General architecture of the BFO concept

8.6.1 Non-fixed Number of Stations

The main steps of this concept can be summarised in the following points:

1. *Set a desired cycle time C.*
2. *Set the maximum peak time to (cycle time $\times$ var), where var $\in [1, 2]$.*
3. *Balance the line (see Chapter 6).*
4. *If satisfying balancing, then continue; else return to 2.*
5. *Test the corresponding AL using the OGA. Evaluate the efficiency of the corresponding line (makespan and idle time). Is the effective cycle time CE close to the desired one?*
6. *If satisfying solution, then assembly line architecture for the family of products found; else return to 1 to try another desired cycle time C.*

The method has been tested on randomly generated problems. The operation durations were generated randomly according to a continuous distribution in the range [5..100]. The number of variant operations is generated uniformly in the range [10 .. $(number\,of\,operations)/3$]. An operation is called variant if its duration is null for at least one variant of the product family. For each operation, the number of precedence constraints is generated randomly in the range [0..8]. The percentage of production requirements is the same for all the variants. The program was executed for a number of operations varying from 50 to 500 and for a number of variants varying from 1 (single product) to 505. For each instance, the optimal number of stations N_{opt} is known. The stopping criterion for the balancing is attained when the number of stations N is equal to N_{opt}. The program was executed more than 25 times (for each instance of the problem) and the optimum solution was found every time. It takes less than a minute for small size instances and less than 2 minutes for large size instances (tests were done on a Pentium II 333 MHz). The method use a population of 50 individuals. As the optimal solution of the ordering problem which results from the balancing is unknown, the stopping criterion for the ordering was fixed to 5 minutes. Table 8.1 presents the results obtained. A set of instances where the maximum peak time is less than the cycle time was allowed to explore the search space. The corresponding solutions are characterised by a high number of stations and reduced makespan.

Note that, as the mix will change with the consumer's demand, it is important to simulate several mix for a given AL.[6] The designer will choose a line that yields similar results for different mixes. To obtain a feasible MPAL, the variants (models) must have some similarities. If the models are significantly different, it is difficult for stations to cope with the differences during assembly. Also, if the set-up time is not negligible, the strategy of changing

[6] The simulated mix has to be as close as possible to the future mix generated by the consumer's demand.

Table 8.1. Results of multi-product AL (BFO)

CT	MPT	MS	SN
7	7	149	8
7	8	167	7
7	9	172	6
7	10	172	6
7	11	172	6
7	12	172	6
7	14	192	6
7	16	192	6
7	20	169	6

CT	MPT	MS	SN
8	7	149	8
8	8	180	7
8	9	179	6
8	10	191	5
8	11	191	5
8	12	191	5
8	14	191	5
8	16	191	5
8	20	208	5

CT	MPT	MS	SN
9	7	149	8
9	8	180	7
9	9	203	5
9	10	199	5
9	11	199	5
9	12	199	5
9	14	227	4
9	16	227	4
9	20	227	4

CT	MPT	MS	SN
10	7	149	8
10	8	180	7
10	9	203	5
10	10	218	5
10	11	216	5
10	12	218	4
10	14	218	4
10	20	218	4

CT	MPT	MS	SN
11	7	149	8
11	8	180	7
11	9	203	5
11	10	218	5
11	11	241	4
11	12	243	4
11	14	248	4
11	20	248	4

CT	MPT	MS	SN
12	7	149	8
12	8	180	7
12	9	203	5
12	10	218	5
12	11	241	4
12	12	243	4
12	14	264	4
12	16	264	4

CT	MPT	MS	SN
13	7	149	8
13	8	180	7
13	9	203	5
13	10	218	5
13	11	241	4
13	12	243	4
13	13	273	3
13	14	281	3
13	16	281	3

CT	MPT	MS	SN
14	7	149	8
14	8	180	7
14	9	203	5
14	10	218	5
14	11	241	4
14	12	243	4
14	14	298	3
14	16	291	3
14	20	303	3

CT : Cycle time
MPT : Max peak time
MS : Makespan
SN : station number

variants must be discarded. The main factors influencing the successful design and efficient operation of a MPAL are the following:

- The line length increases as balancing efficiency decreases.
- The greater the number of stations the easier it is to find a good scheduling.
- A small number of stations yield a high ratio of reliability and a high idle time.
- The complexity of sequencing increases with the number of variants.
- Sensibility to production demands deviation for each variant.
- Process time deviation due to variants of each station (use of maximum peak time parameter).
- Operator task time variation (use of stochastic duration time).

- The optimum depends on the cycle time and maximum peak time (Figure 8.4). It corresponds to the solution having the minimum number of stations and the minimum makspan.
- The use of simulation algorithms to validate results.
- The use of buffers to solve starving and blocking problems of the line.
- Take into account set-up time and operator moves to evaluate process time of stations.

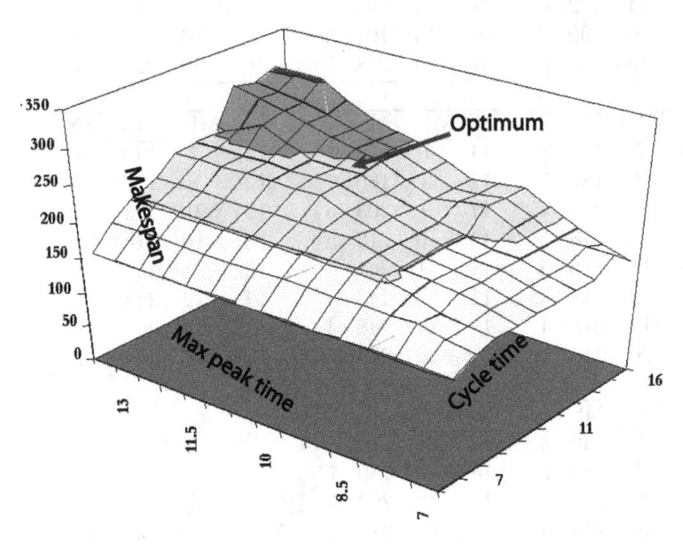

Figure 8.4. Distribution of makespan versus cycle time and maximum peak time

8.6.2 Fixed Number of Stations

The main features of the approach are presented below.

1. Set preferences of the 'variant process time standard deviation' and of the 'station misbalance' (see Chapter 6).
2. Balance the line (see Chapter 6).
3. If the balancing is satisfied, then continue; else return to 1.
4. Test the corresponding AL using an OGA. Evaluate efficiency of the corresponding line.
5. If the solution is satisfied, then AL architecture for the family of products found; else return to 1 and try other preferences.

The objective of the method is to balance the *average* workload among the stations and the process among the variants. In this case the scheduling module serves as a validation technique of the balancing module. Some results of the proposed approach will be detailed in Chapter 10.

Part IV

The Integrated Method

9

Evolving to Integrate Logical and Physical Layout of Assembly Lines

9.1 Introduction

ALD is well known as the elaboration of the LL and the PL of the line. The LL consists of the distribution of tasks among stations along the line, while the PL decides about the disposition of some variants (*e.g.* the stations, conveyor(s), *etc.*) on the shop floor. The goal of most approaches consists of the equalisation of the workload of stations to the cycle time or the minimisation of the number of stations, whereas other factors (such as traffic problems, station congestion, transport network, *etc.*) may also heavily affect the system. A new method is proposed for an LL taking the topology of the line (facility) into account. This architecture represents a rough idea of the PL of the future line. Background and motivations of the approach presented are briefly described in Section 9.2, while the AL layout problem is presented in Section 9.3. The concentration is focused on the utility of the *workcentres clustering* phase and the benefits of the proposed architecture are fully explained. The integrated approach is presented in Section 9.4, where the interactive and the optimisation phases are detailed. Results of the approach on an industrial case study are presented and discussed in Section 9.5.

9.2 The State of the Art

Several studies have been published about facilities planning [7], [51], [158] and [164]. However, bridging the gap between the LL and PL is completely neglected. The authors also tackled the cell formation problem in various ways [35], [84] and [97], but these approaches are more focused on cellular manufacturing (CM), group technology (GT), and material flows, and are not able to deal with the LL. A global approach which was a result of the SCOPES project [39] considers the main factors that affect the performances of the AL. The PL module, which is based on a simulation package, is executed after the LL. Lucertini *et al.* [89] presented a unified framework for designing production plant and its corresponding network of material flow. For more information, the reader is referred to [2, 66, 77, 81]. Different philosophies of layout are appropriate for different manufacturing environments:

Fixed Position Layouts. Some products are too big to be removed, so that the product remains fixed and the layout is based on the product size and shape (*e.g.*, airplanes, ships and rockets).

Product Layouts. The product layout is typically of high-volume standardised production. An AL is product layout, because assembly facilities are organised according to the sequence of steps required to produce the item. Product layouts are desirable for flow-type mass production.

Process Layouts. Process layouts are mostly effective when there is a wide variation in the product mix. Each product has a different routing sequence associated with it. Process layouts have the advantage of minimising machine idle time.

GT Layouts. The GT concept seems to be best suited for large firms that produce a wide variety of parts in moderate to high volume. The GT layouts are product family oriented, while the process layouts are machines functions oriented.

9.3 Assembly Line Design

The main idea behind the design of ALs is that, for complex products, the assembly system must be decomposed into subsystems which are easier to manage than the entire one. The line is decomposed into several linked sub-lines (called workcentres in the remainder of this chapter), with their own cycle time, reliability, and station requirements. Each sub-line is attributed to one or many sub-assemblies. The routing of a product from one workcentre to another is fixed according to a line flow topology. The main topology of the line is not necessarily a linear one. With classical line balancing techniques, a way to tackle the line balancing problem would be to balance each workcentre separately. But in real conditions, some operations allocated to a given workcentre could be affected by another one and linked to the former.

9.4 Integrated Approach

Several attempts have been made in the field of assembly to give assembly workshops a general structure and identical to that of machining systems. ALs still retain a linear structure due to the supply, high robustness, and ease of management. The drawbacks may be poor fault tolerance and routing flexibility [2]. The main task of the proposed line layout *integrated method* is to cluster twice the tasks (two levels)(Figure 9.1).

1. workcentre clustering: partition a set of tasks performing alike activities together. This leads to a number of workcentres.
2. Station clustering: assign tasks to stations. This leads to a number of stations in each workcentre.

This assignment has to take into account precedence, transportation, and synchronisation of the sub-assemblies in order to find the best value of ratio between clustering and transportation index. The second phase permits the design of a workcentre dealing with objectives like workload balancing, cost, reliability, imbalance between variants, *etc.* The problem is composed of three interdependant sub-problems: workcentre clustering, station clustering, and workcentre synchronisation [126, 134].

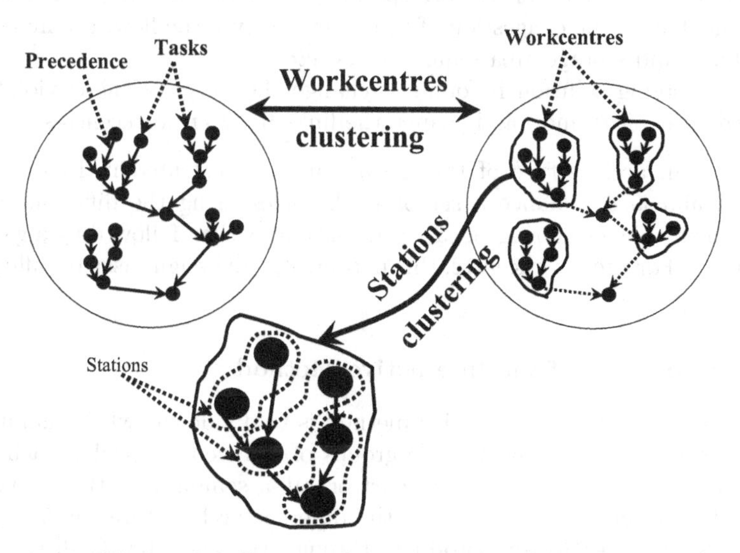

Figure 9.1. Integrated approach of the line layout

The results obtained using the balancing module permit one to know the distribution of tasks and resources along the AL. The PL module determines the space requirements taking into account congestion and material storage, handling systems, and so on (Figure 9.2). The whole methodology can be described as follows:

- Set the desired workcentres, and for each of them assign tasks into workcentres, dealing with precedence graphs, set the desired number of stations, and set the desired cycle time.
- Set the desired links between workcentres.

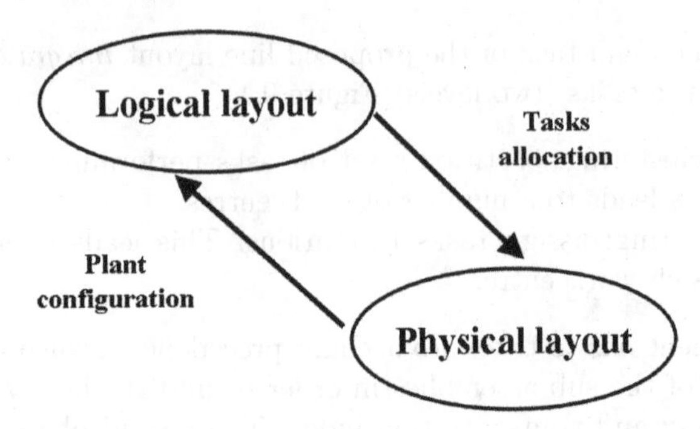

Figure 9.2. LL and PL interaction

- Balance the whole plant (set of workcentres).
- Position workcentres and stations.
- Evaluate the efficiency of the corresponding plant layout using a simulation package. Check the congestion of the plant, analyse the flow, the material handling, and storage area requirements, *etc.*
- If no satisfying solution is found, exchange the tasks (without violating precedence constraints) and change the links between workcentres.

The overall architecture of the LL module is illustrated in Figure 9.3. The main aim is to balance a set of workcentres using the different links between them. The clustering (local optimisation) is then followed by a global design phase. For each workcentre, this permits one to assign tasks to different stations.

9.4.1 Development of the Interactive Method

The principal goal of the interactive method is to divide the whole manufacturing facility into small manageable groups of workcentres (cells), each cell being dedicated to a specified set of part or sub-assemblies. In the following sections the concentration will be on the different architectures of ALs and on the flow of materials and products through the assembly facility. Various strategies for organising resources are described and some techniques to help designers to interact[1] with the system are presented. Simple indexes to evaluate the performance of these configurations are discussed.

Workcentre Clustering

The aim of this phase is to cluster tasks among workcentres. In Figure 9.4 the left top (a) represents the precedence graph of product, while the right top

[1] It makes no sense to ask the computer to find a solution for something which is obvious, *e.g.* the tasks that the human can do better. It is also a waste of time to solve problems for which the computer needs a lot of data to find simple solutions while the human can find them without any difficulty.

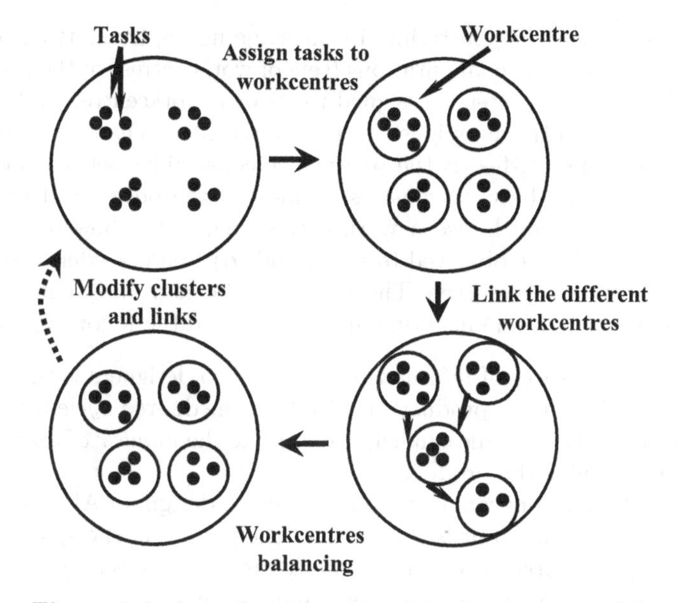

Figure 9.3. Overall architecture of the line layout module

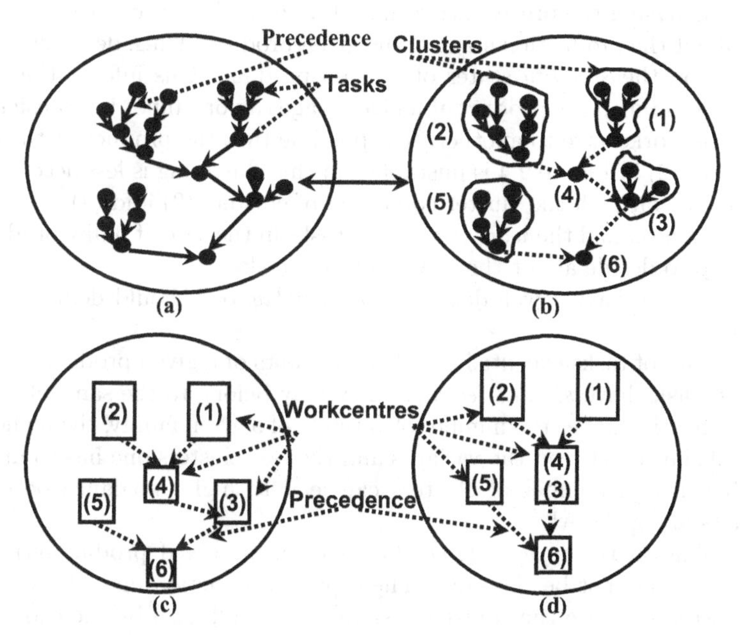

Figure 9.4. workcentre clustering phase

(b) represents one possible clustering. In the same figure, the bottom (c) and (d) represent two possible configurations (sets of workcentres) of the proposed clustering. In (c) each cluster is assigned to its own workcentre, while in (d) the clusters (3) and (4) are assigned to one workcentre. The only hard constraint that must be satisfied is the precedence constraints between clusters. The precedence graph between clusters decides on the position of workcentres. In contrast, the arc between workcentres defines the flow of products between them. As can be observed from (c) and (d), the precedence between the different clusters is preserved. The criteria and constraints that influence the choice of a given graph clustering are described in the following points:

- One of the first pieces of information provided to designers is the desired throughput of the given product; this leads to the desired *cycle time* of the line. The desired cycle time can help to estimate the *number of workcentres* needed to assemble the product.
- Generally, designers never start from scratch to design an AL. One of the most important constraint is the space of the plant and the space of each sub-plant, each workcentre, and so on. Thus, the *number of stations* of each workcentre is more-or-less known in advance.
- Since the line must operate according to a line flow topology, only clusters that can satisfy the *precedence constraints* between tasks are valid.
- When analysing the precedence graph of a given cluster, one can have an idea about the production stage of the given product. Thus, designers have an idea on the *stability* states of a given product. This information will help in deciding if the product at a given stage can or cannot be transferred from one workcentre to another. It is possible that the product at the end of cluster (2) in Figure 9.4 is unstable and this clustering is less acceptable in comparison with the cluster composed of clusters (2) and (4).
- The *work level* and the *work position* of tasks in the case of bulky products can help to decide about the way to cluster tasks.
- Each time we have a well-defined *sub-assembly*, one should dedicate it a cluster.
- Given a set of tasks executed on all the variants of a given product family, if these tasks have *similar features*, they may belong to the same cluster.
- Generally, the higher the number of variants of a given family, the higher is the imbalance between the variants and the less clusters one has to make. Making a high number of clusters can lead to high imbalance between variants along the AL.
- Depending on the *type of production* (batch or mixed production), the clustering may not be the same. The type of production may also change the clustering, since the transfer system may be affected by the choice.

On the other hand, the main parameters that influence the workcentre clustering can be summarised as follows:

- The importance of the human is often disregarded while evaluating the AL performance. In order to deal with human behaviour, a close interaction between designers and workers can define useful clusters that satisfy workers desires and enhance job quality.
- One of the basic pieces of information to the clustering phase is 'how far geographically the different workcentres are'. Indeed, the transfer system depends on the *distance* between the workcentres.
- Components *storage* space is one of the hard constraints in AL designs. Since each assembly task is linked to a given component, it is quite easy to detect if the storage space needed for a given set of tasks exceeds the storage space of a given workcentre.
- The *feeding* system of the different components can help to decide about the grouping or not of a set of tasks. In the case of the *'kiting'* philosophy, the feeding has only a minor influence on the choice of the clustering.
- The plant layout, its obstacles (walls, paths), the specific stations (quality control cells, and painting stations, *etc.*) may introduce constraints on the position of workcentres and their links.
- The number of *operators* permits one to define the number of workcentres.
- Indeed, it makes no sense to introduce a transfer system between two workcentres, as each one contains only one task. Techniques like 'cell formation' and flow matrix can help to decide about the acceptance or not of a proposed clustering.

It is important to note that the results of this phase constitute a *local* optimisation of the line layout problem. Indeed, this clustering permits one to narrow the search space, whereas the results of the LL module constitute a *global* optimisation.

Workcentre Links

Some workcentres may serve to assemble a subassembly which is injected as a whole in the main line. Some stations, like packaging, which may be used for several products in the same facility, are at the confluence of two or more ALs. Thus, different lines or workcentres are linked to yield several line *topologies*, as illustrated in Figure 9.5. Four workcentres are linked to a main line according to a 'fishbone' topology, and the main line separates into two others at its end. Different links (if they exist) represent just a *logical* link between workcentres, as shown in the example of Figure 9.5 (the workcentre (4) is linked to the station (5) of the main workcentre). This means that the transfer system has to put the product leaving workcentre (4) on station (5) of the main workcentre.

There are two general kinds of link: links with operations exchange and links without operations exchange. The possible links between workcentres are described in the following sections.

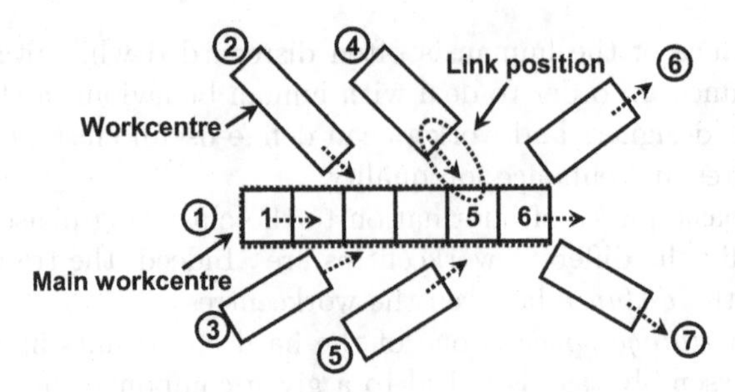

Figure 9.5. Example of plant topology

Link Without Operation Exchange. A simple link is when two workcentres are linked logically without any exchange of tasks. Such links only help to decide about the flow among workcentres. There are three possible configurations (Figure 9.6). The arrows represent just the flow of the product inside the workcentre.

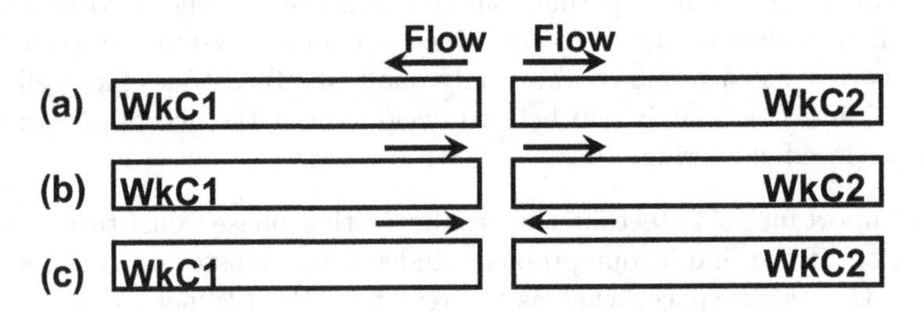

Figure 9.6. Possible links between workcentres

- Physically, the two workcentres may be put in parallel; this means that the two sub-assemblies start at the first station of each workcentre.
- The last station of workcentre WkC1 is linked to the first station of workcentre WkC2. Once WkC1 finishes its work on the product, it transfers it to WkC2.
- The last station of workcentre WkC1 is linked to the last station of workcentre WkC2. In this case there are two possibilities.
  1. WkC1 finishes its work on the product and transfers it to WkC2, or WkC2 finishes its work on the product and transfers it to WkC1.
  2. Suppose there is another workcentre WkC3 connecting the two workcentres. Thus, the sub-assemblies assembled on workcentres WkC1 and WkC2 are transferred into WkC3. Then, the two sub-assemblies may be assembled together on WkC3.

Sharing Stations. The second kinds of link correspond to a set of workcentres *sharing* physically one station (see Figure 9.7). The product passing through the different workcentres has to visit the shared station. This kind of station can be found in the following situations:

- In the contact point of many parallel workcentres. Suppose there exist a set of tasks done by a robot. Note, that the cost of a robot is generally high and it is more beneficial to share it to execute the same task relative to the different workcentre. For two 'paced' workcentres, the process time of the shared station may not exceed half of the cycle time.
- The contact point belongs to the main workcentre. Each workcentre assembles a sub-assembly relative to a variant and the main workcentre integrates the different sub-assemblies to the main product.

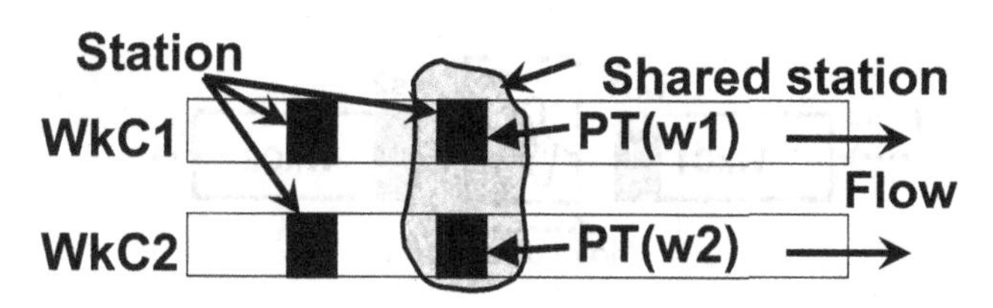

Figure 9.7. workcentres sharing station

The hard condition to share an operator between many stations is that

$$\sum_{w=1..NbLinks} PT(w) \leq Minimum(CycleTime) \tag{9.1}$$

where $PT(w)$ is the process time of the shared station on the product passing through workcentre w, $NbLinks$ is the number of workcentres sharing this station, and $Minimum(CycleTime)$ is the cycle time corresponding to the fastest workcentre.

For each cycle time, the shared station has to work on the products relative to the different workcentres. This means that on each period equal to the cycle time this station has to handle each of the $NbLinks$ products. Suppose that the process time of the given station relative to each workcentre is equal to the minimum cycle time, then the process time of the station must be less than or equal to the corresponding cycle time. By the way, the *sum* of the process times corresponding to the different workcentres must be less than or equal to the minimum cycle time. The *upper bound* is then $Minimum(CycleTime)$. Thus, the theoretical maximum process time of the shared station must verify the inequality in Equation (9.1). This upper bound is relative to the synchronised model: the station always begins with the product passing through the fast workcentre (corresponding to the minimum cycle time). Many other combinations (synchronisation) are possible, especially in the case of MPALs [146].

Link With Operator Move. The third kind of link corresponds to the case where two physical stations of two different workcentres belong to the same logical station (Figure 9.8). Logically, there is only one station, but physically one part of the job is done on WkC1 and the rest of the job is done on WkC2. One operator (machine, or robot) is used to work on station ws1 and to transport the product from WkC1 to WkC2 (station ws2). The latter continues assembling the product. This can be the case if a heavy equipment Eqp1 is installed on ws1 and another Eqp2 is installed on ws2, and the product has to go successively from WkC1 to WkC2. The main condition to do such an allotment is that the process time on the two stations must verify the following assignment:

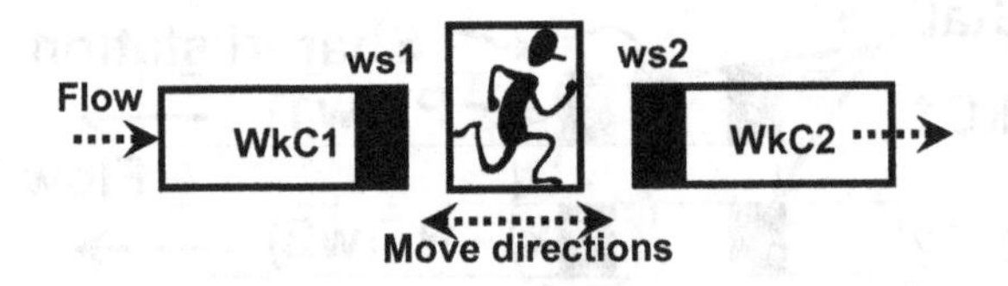

Figure 9.8. workcentre link with operator move

$$PT(ws1) + PT(ws2) + 2 \times Mvt < Minimum(CycleTime) \qquad (9.2)$$

where $PT(w)$ is the process time of the station w, Mvt is the duration of the movement between the two workcentres, and $Minimum(CycleTime)$ is the cycle time corresponding to the fast workcentre.

Note that the same product passes through the two workcentres: the operator (or robot) transfers the product from the first workcentre to the second. Thus, the flow of the two workcentres must be the same, otherwise there is no need for such configuration.

Link With Operations Exchange. Finally, the most interesting kinds of link between workcentres are those where tasks are exchanged among workcentres. This exchange can help to balance the workload of two adjacent workcentres if the *surplus of process time* on one station is transferred to its neighbour. Note that the exchange of tasks is done in only one direction, not in both. The surplus of process time on the *overloaded* workcentre is transferred into the other. Otherwise, the product has to be transferred twice between the two workcentres. Figure 9.9 represents two linked workcentres which are able to exchange tasks between the linked stations: the *second* station of workcentre A and *third* station of workcentre B.

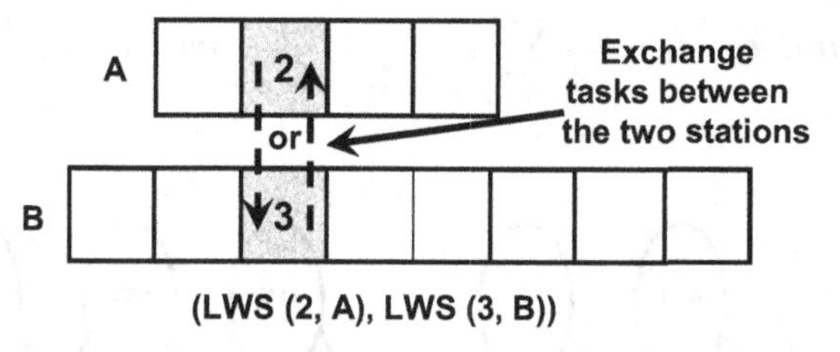

Figure 9.9. workcentre links with tasks exchange

Note, that the links are not mandatory and a workcentre may be isolated from the rest of the line.

9.4.2 Global Search Phase

Figure 9.10 illustrates the input data of this module which helps to balance locally a given workcentre (using only the tasks that belong to this workcentre). The balancing of the line is done using the EPAL heuristic which was introduced in Chapter 6. In order to take advantage of the links between stations, another heuristic has been developed. The 'link node' is the set of stations by which a set of workcentres is linked. For instance, suppose that the link (end(WkC1), end(WkC2)) has been set (*i.e.* the end of WkC1 is linked to the end of WkC2), the link node will be the last station of each workcentre.

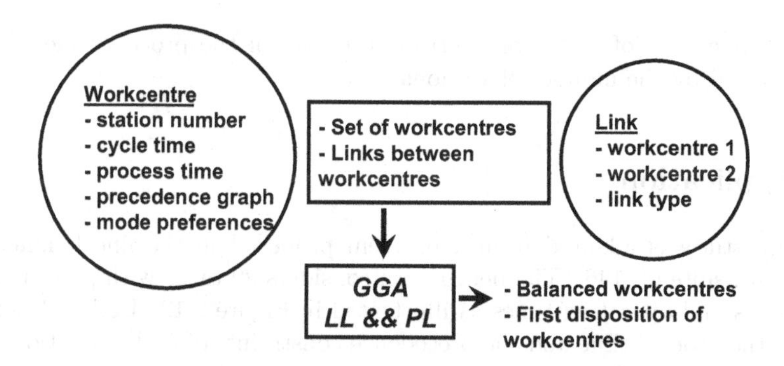

Figure 9.10. Input data of the problem

The two stations in the link node are chosen and all possible exchanges between them (which do not violate precedence constraints and cycle time) are executed, as shown in Figure 9.11. These kinds of move permit one to balance two adjacent workcentres by exchanging tasks between them. The objective is to equalise station durations under a fixed number of station constraints. The following cost function is adopted:

$$minimise \quad f_{EP} = \sum_{j=1..w} \left(\sum_{i=1..N_j} (fill_i - cycletime_j)^2 \right) \tag{9.3}$$

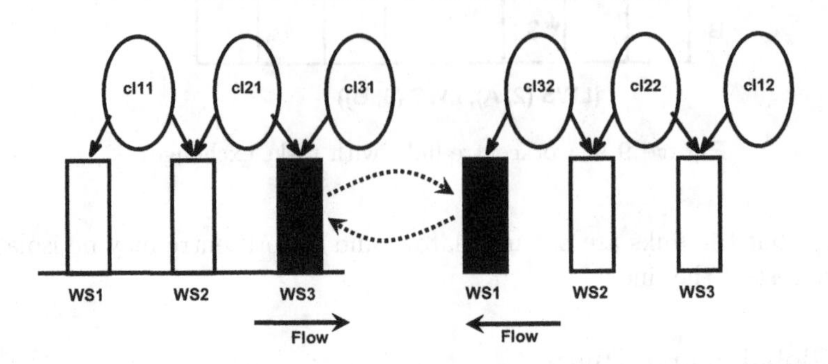

Figure 9.11. Linked wheels heuristic

In other words, for each workcentre, this function minimises the square of the difference between the workload of stations and the desired cycle time, where w is the number of workcentres, N_j is the number of stations of each workcentre, $fill_i$ is the sum of working times on station i, and $cycletime_j$ is the ideal cycle time of workcentre j. This can defined as follows:

$$cycletime_j = \frac{\sum_{(j=1..nbop_j)} time_i}{N_j} \tag{9.4}$$

The cycle time of each workcentre is the sum of the process time of its tasks divided by the number of stations.

9.5 Application

The case study is adapted from a problem proposed in the line balancing benchmark suite of [146]. The benchmark considers 29 tasks with precedence constraints and operating times as illustrated in Figure 9.12. Table 9.1 summarises the process time and the precedence constraints of each operation, as well as their prefered workcentre. We decide to create two workcentres, with the link (end(WkC_A), begin(WkC_B)).

First, the two workcentres are balanced without using any link. Table 9.2 presents a set of solutions for a given number of stations without cycle time restriction and according to an equal piles strategy. NbS_X denotes the desired number of stations, while WkC_X represents the process time of stations for workcentre X. Finally, the link represents the station by which the two workcentres are connected.

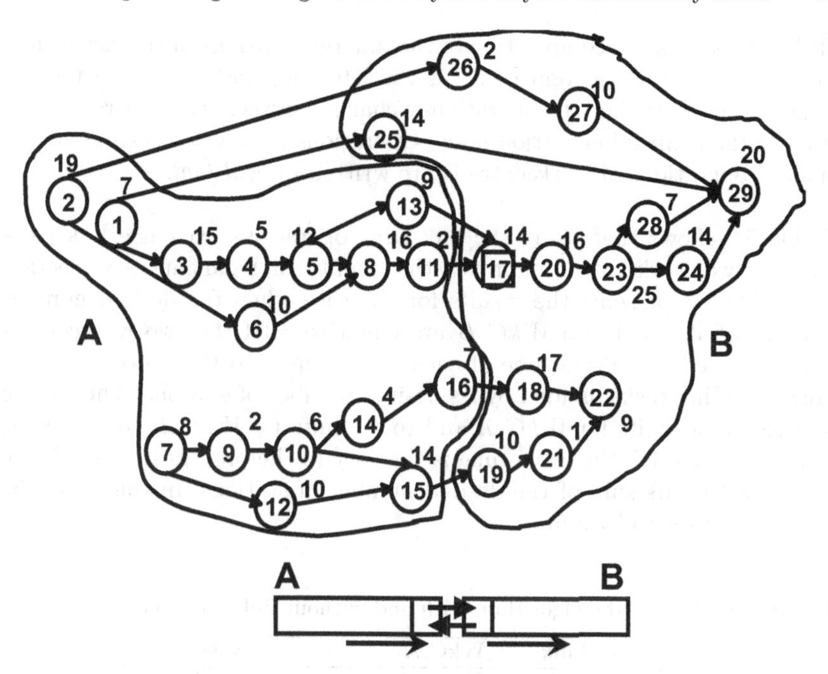

Figure 9.12. Precedence graph of the problem

Table 9.1. workcentre, duration and precedence constraints of each task

Op	WkC	Duration	Preds	Op	WkC	Duration	Preds
1	A	7		**16**	A	7	8, 14
2	A	19		**17**	B	14	11, 13
3	A	15	1	**18**	B	17	16
4	A	5	3	**19**	B	10	15
5	A	12	4	**20**	B	16	17
6	A	10	2	**21**	B	1	19
7	A	8		**22**	B	9	18, 21
8	A	16	5, 6	**23**	B	25	20, 22
9	A	2	7	**24**	B	14	23
10	A	6	9	**25**	B	14	1, 7
11	A	21	8	**26**	B	2	2
12	A	10	7	**27**	B	10	26
13	A	9	5	**28**	B	7	23
14	A	4	10	**29**	B	20	24, 25, 27, 28
15	A	14	10, 12				

Table 9.3 shows the composition of the different stations in the case where the two workcentres are connected (at the left-side) and not connected (at the right-side). Note that the operation exchange between workcentres is only allowed at the connection station node. Operations from workcentre A mixed with some operations of workcentre B are written in *bold font.*

Table 9.2 shows that using the link (*i.e.* 'operations exchange link') between the two workcentres improves the quality of balancing (see Section 9.4.1). Table 9.3 presents the results for the case when the desired number of stations of WkC_A and WkC_B are equal to 3. If the two workcentres are disconnected, the cycle time of each one is equal to the process time of its corresponding tasks divided by its desired number of stations. Thus, cycle time is set to 60 units for WkC_A and to 49 units for WkC_B. In contrast, if they are connected, the cycle time is then set to the process time of all the tasks divided by its sum of the desired number of stations. In this case, the cycle time was set to 54 units.

Table 9.2. Results of the algorithm, with and without links between workcentres

(NbS_A, NbS_B)	Link	WkC_A	WkC_B
(3, 3)		61, 60, 58	49, 48, 48
(3, 3)	(3, 1)	56, 54, 52	54, 55, 53
(4, 3)		39, 46, 45, 49	49, 48, 48
(4, 3)	(4, 1)	47, 46, 47, 42	46, 48, 48
(4, 4)		43, 47, 45, 44	27, 38, 39, 41
(4, 4)	(4, 1)	41, 43, 41, 43	38, 38, 39, 41
(5, 3)		35, 37, 36, 37, 34	49, 48, 48
(5, 3)	(5, 1)	38, 44, 42, 41, 37	40, 41, 41
(5, 4)		35, 37, 36, 37, 34	27, 38, 39, 41
(5, 4)	(5, 1)	37, 35, 36, 36, 31	33, 36, 39, 41
(6, 3)		30, 34, 30, 30, 30, 25	49, 48, 48
(6, 3)	(6, 1)	35, 37, 36, 37, 34, 37	28, 39, 41

Table 9.3. Process time and list of tasks of each station with (NbS_A=3, NbS_B=3)

WkC	PT	Ops (without link)	PT	Ops (with link)
A	61	0, 2, 6, 8, 1, 5	56	0, 6, 1, 11, 5, 8
A	60	9, 11, 13, 3, 14, 4, 12	54	9, 2, 3, 4, 7
A	58	7, 10, 16, 15	52	13, 15, 14, **17**, **18**
B	49	17, 20, 21, 18, 25, 26	54	**12**, **10**, 20, 21, **16**
B	48	19, 22, 27	55	22, 19, 23
B	48	23, 24, 28	53	25, 24, 28, 26, 27

Table 9.4 shows that the balancing obtained using the link between the two workcentres (connected by the fourth station of WkC_A and the first station of WkC_B) is better than the first one. The results show that the links allow to smooth the workload of the different stations along the two workcentres. Indeed, the maximum difference is not more than 6 (48 – 42) in the second case (with link), while it is equal to 10 (49 – 39) in the first case (without link).

Table 9.4. Process time and list of tasks of each station with (NbS_A=4, NbS_B=3)

WkC	PT	Ops (without link)	PT	Ops (with link)
A	39	0, 2, 3, 4	47	0, 2, 3, 6, 4
A	46	1, 12, 6, 5	46	8, 9, 11, 12, 1
A	45	7, 8, 11, 9, 13, 15	47	5, 7, 10
A	49	10, 16, 14	42	13, 16, 15, **17**
B	49	17, 18, 20, 21, 25, 26	46	18, 20, 21, 25, 26, **14**
B	48	19, 22, 27	48	19, 22, 27
B	48	23, 24, 28	48	23, 24, 28

10

Concurrent Approach to Design Assembly Lines

10.1 Introduction

The integrated method is composed of three independent modules, namely the PA, the OMT and the LL. Designers always have an idea on how to design the product. At the same time, the proposed line designs are most of the time influenced by the way the product will be assembled (PA and OMT outputs). In fact, there is a *simultaneous* design of the product and its line, such that fixing the product structure limits the possibilities of the line architecture. This chapter is organised as follows. In Section 10.2, a brief description of the ALD problem and its constraints and objectives are discussed. The integrated approach is introduced in Section 10.3. Results of two industrial case studies are presented in Section 10.4.

10.2 Concurrent Approach

The principal features of the line layout module were introduced in previous chapters. That is, the LLs for ALB and RP were presented respectively in Chapters 6 and 7. The balance for operation concept was presented in Chapter 8.

The following steps are adopted when executing ALD: (1) fix on the assembly type, (2) draw the precedence graph, (3) describe the whole process, (4) decide about the line speed, the cycle time and number of stations, (5) use the integrated approach, and finally (6) evaluate the efficiency of the line obtained (see Figure 10.1). If the proposed design is not satisfactory, the operating modes or precedence graph will be modified. Product modifications can also be envisaged at a late stage, but only if other modifications are ineffective. A feedback to previous steps of the process is always possible. In the next section, a framework of the proposed AL layout is presented.

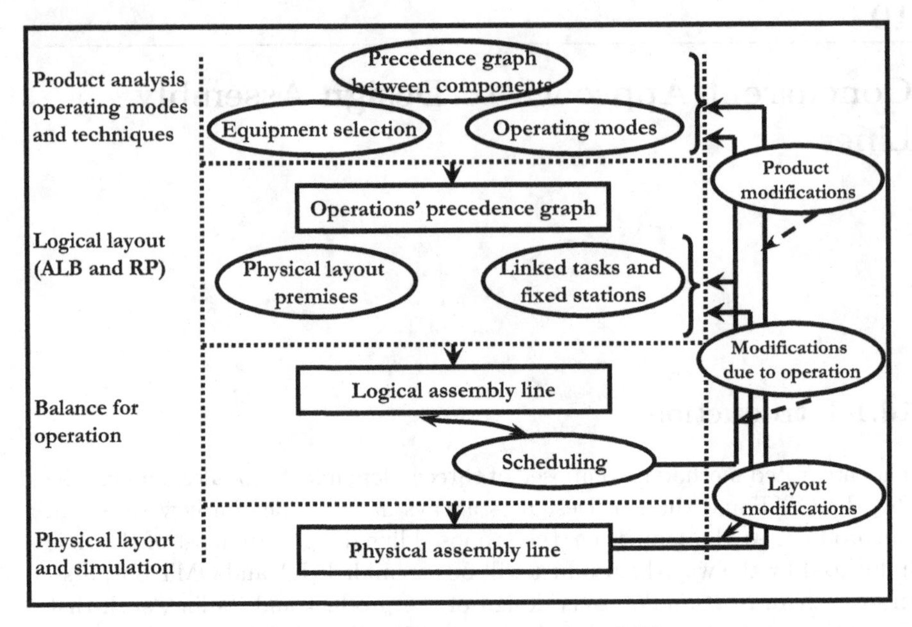

Figure 10.1. Concurrent design of AL

Thus, the integrated ALD approach is illustrated in the following pseudo-code:

repeat
Fix the operating modes and techniques of tasks;
Select the equipments for each task;
Cluster tasks between workcentres;
Impose grouped tasks and fixed stations to recover existing stations;
Balance for operation the line;
  repeat
    Propose the LL of the AL;
    Test the operating efficiency of the AL;
  until *satisfactory solution has been found;*
  Decide about the disposition of the stations, conveyors, buffers, on the shop floor;
  Simulate the AL to investigate the impact of architecture obtained;
until *a satisfactory AL architecture has been found;*

10.3 Assembly Line Design

The proposed method is built upon many collaborations with industry [124]. Its main steps can be summarised as follows (see Figure 10.2):

- **Preparation**. The designer introduces the input data (tasks, resources, constraints, preferences, *etc.*);
- **Optimisation**. The optimisation method proposes a line architecture (stations contents, their order, *etc.*);
- **Mapping**. This allows the designer to analyse and test the results using a simulation package.

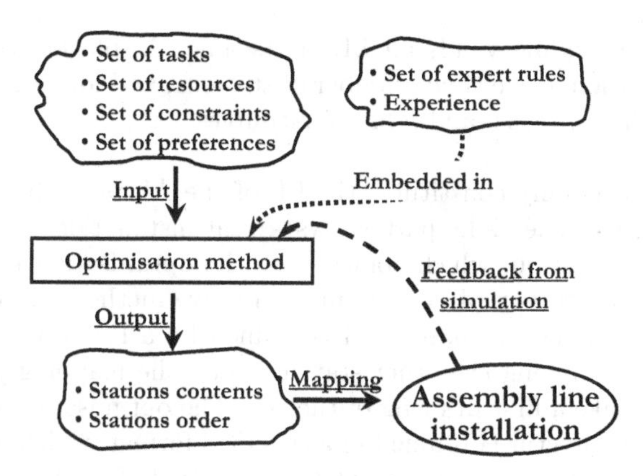

Figure 10.2. Design method

10.3.1 Data Preparation Phase

Once the product and the existing resources of the enterprise have been analysed, a set of assembly plans is proposed, as well as their preferable resources [110]. The method gives the following input for the optimisation phase, as illustrated in Figure 10.3: the desired number of stations, the desired cycle time, and for each task the precedence constraints and the user's mode preferences.

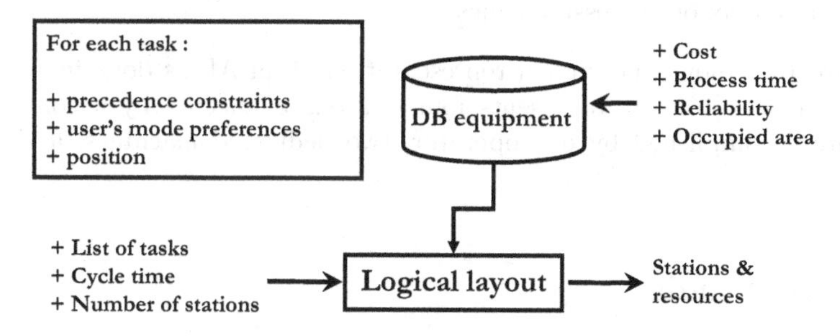

Figure 10.3. Data flow of the ALD method

10.3.2 Optimisation Phase

This phase constitutes the evolutionary computation part of the methodology. The approach is based on GAs, and many industrial designers' ideas are embedded in the method as heuristics.

10.3.3 Mapping Phase

The optimisation module yields an LL of the line. A solution contains the following information: cycle time, number of stations, and for each station the process time, a list of tasks, and a list of resources.

This information only constitutes the LL of the AL presented on the left side of Figure 10.4. The right part shows a real installation of an AL and its relative representation, which comes from the optimisation module. The missing step of the PL is replaced by an interactive method. Each station is represented by an object (square) and is defined by a list of tasks, a list of resources, its order among the other stations, *etc.* The mapping phase helps the designer to make a first drawing of the AL. The optimisation module has to save the AL architecture obtained in a specified format, which is then used by the simulation software package (AUTOMOD) [169]. The AUTOMOD input data needed to design the system are:

- locations – they correspond to real location of stations;
- entities – they represent pallets, parts, and all items which are moving between different locations;
- resources – they describe operators, conveyors, machines which are able to move entities between two locations, or can execute an operation;
- tasks – the description of tasks as well their order on each station;
- path network – this consists of paths followed by resources and/or entities in the real system;
- processing – this allows one to define possible destinations of an entity leaving a location. The station's duration spent on each product is deduced using a matrix of processing times.

Figure 10.5 shows the virtual representation of an AL as done in AUTOMOD software [169]. It represents four stations connected by a conveyor. Tasks are accomplished by one operator, two dedicated machines, and one robot.

10.4 Case Studies

The following sections present the application of the EPAL and RP approaches using two industrial case studies.

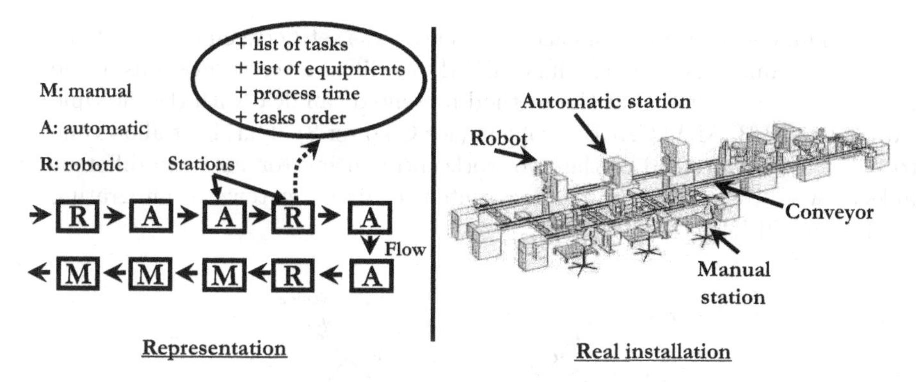

Figure 10.4. Relationship between the real architecture of the line and its representation

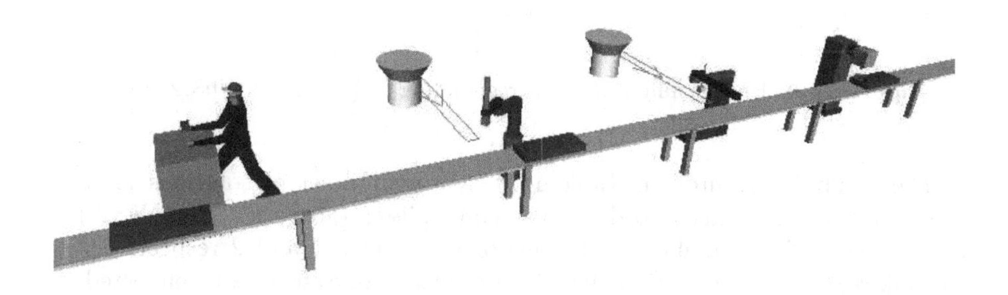

Figure 10.5. An AUTOMOD representation of an AL

10.4.1 Assembly Line Balancing Application: Outboard Motor

The product studied is an 'outboard motor marine' engine. The aim is to balance the workload for a fixed number of stations. The line produces between 9 and 11 engines per hour depending on the period of year. Table 10.1 summarises the tasks performed on the product and shows the workcentre, the process time (tenths of an hour) and the precedence constraints for each task. The tasks (155 in total) whose *number_id* is less than 450 belong to the first workcentre (the remaining ones belong to the second workcentre). The plant in this case is composed of two workcentres to balance the workload of the AL using different numbers of stations. The first configuration assumes that there is no link between workcentres, while in the second one the workcentres are linked by their last stations. The two workcentres are linked by an 'operations exchange' link (for more details see Chapter 9).

The number of stations of each workcentre was set between one and four. Table 10.2 summarises the results of EPAL for different configurations of the AL. These results show that the method presented can deal with the multiple workcentre ALB (MWkCALB) problem (see Chapter 3). That is, it allows one to balance the workload of the two workcentres using (or not) the different links between them. The approach presented is a first step towards integrating the LL and PL of ALs.

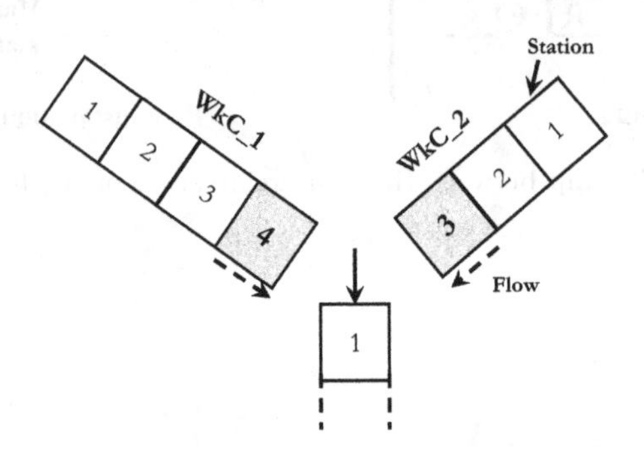

Figure 10.6. Plant configuration corresponding to (NbS_1=4, NbS_2=3)

The results, presented in Table 10.2, represent a set of solutions for a given number of stations according to an equal piles strategy. (NbS_1, NbS_2) denotes the desired number of stations of workcentre 1 and 2 respectively. The link represents the stations by which the two workcentres are connected. WkC_1 (or WkC_2) represents the process time of stations for workcentre 1 (or 2). Here, the results corresponding to $(NbS\_1 = 4, NbS\_2 = 3)$ are given. The results show that by adding the link between the two workcentres the whole line may be better balanced. For instance, the stations workloads obtained in the case of $(NbS\_1 = 4, NbS\_2 = 3)$ are

- without link $\{WkC\_1 : 82, 82, 82, 82$ and $WkC\_2 : 73, 68, 58\}$
- with link $\{WkC\_1 : 75, 76, 75, 75$ and $WkC\_2 : 76, 76, 74\}$.

Table 10.3 shows the results obtained for the two workcentres, where the number of stations is set to four for the first workcentre (and three for the second). This table shows for each workcentre the process times of the different stations and their corresponding tasks. The first solution corresponds to the case where the two workcentres are disconnected. The second workcentre is badly balanced because of the hard precedence constraints between tasks and the process time of the different tasks attributed to this workcentre. The architecture where the two workcentres are connected yields the second solution. The following tasks $\{127, 420, 427, 425, 292, 237, 395, 400, 402, 405\}$ were transferred from the first workcentre to the second one. The second solution is better balanced over the two workcentres. These results show that taking into

Table 10.1. workcentre, duration and precedence constraints of operations

Op	WkC	Duration	Preds	Op	WkC	Duration	Preds
12	0	0.1		**307**	0	3.1	
15	0	0.5		**310**	0	1.2	
17	0	1		**312**	0	3.6	225
20	0	0.5		**317**	0	2.2	
22	0	0.7		**320**	0	1.5	
25	0	0.3	22	**322**	0	3.2	
27	0	3		**325**	0	1.3	
30	0	3.6	27	**327**	0	1.3	
32	0	3.6		**330**	0	1.2	
35	0	0.3		**332**	0	1.2	
37	0	3.5		**335**	0	5	
40	0	3.5		**337**	0	1	
42	0	5.1	40, 35	**338**	0	1.7	
45	0	5.1		**339**	0	4.4	185
47	0	5		**347**	0	3.6	185
50	0	5	40	**350**	0	1.7	
52	0	0.8		**355**	0	4.8	240
55	0	0.6		**357**	0	1.1	310
60	0	2.8		**367**	0	0.6	
62	0	0.9		**372**	0	3.8	355
65	0	1.3	60	**380**	0	6	355
67	0	2.5	62, 65	**387**	0	0.6	
70	0	4.6	62, 65	**395**	0	1.7	357, 339, 312,
75	0	10	62, 65, 70				350, 265, 292,
87	0	1.8	75				237, 243, 247,
90	0	3.3					372, 380, 252,
97	0	1.7					367, 280, 387,
100	0	0.7					282, 338
102	0	5.5		**397**	0	2.2	185
106	0	2.4		**400**	0	6	395, 397, 67
110	0	4.2	106, 122	**402**	0	1	400
112	0	2.4	110	**405**	0	3.6	402
115	0	2.6		**410**	0	4.2	
117	0	4		**415**	0	3	410
120	0	1.1		**417**	0	4.4	
122	0	2.2		**420**	0	4	415
125	0	1.7	120	**425**	0	1.4	415
127	0	0.6	47, 90, 125	**427**	0	5.5	415
130	0	4.4		**430**	0	3.8	417
132	0	4	130	**450**	1	4	420, 425, 427, 430
135	0	1.2		**452**	1	1	
137	0	3.5	135	**455**	1	3.4	
140	0	5.5	137	**457**	1	5.5	455
147	0	5.5		**460**	1	2.5	455
152	0	3.3	147	**462**	1	4.5	457, 460
155	0	3.3	152	**465**	1	10	462
157	0	3.3	155	**467**	1	4.5	
160	0	2.2		**470**	1	7	
162	0	2		**475**	1	4.5	
165	0	1.5	162	**477**	1	5	
167	0	3.3		**480**	1	3.4	
170	0	1		**482**	1	8	
172	0	5.5	170	**485**	1	4.5	
180	0	3.3		**487**	1	2	
182	0	4.2	180	**490**	1	3.3	
185	0	4.4	182	**507**	1	1	
190	0	1.8		**512**	1	5	
193	0	2.5	185	**515**	1	4.8	
297	0	1.6	185	**516**	1	1.2	
225	0	2.4	190, 193, 197	**517**	1	3.3	
227	0	9.5		**520**	1	8.5	
237	0	7.2		**522**	1	5	
240	0	1.5		**525**	1	4	
241	0	2.2	240	**527**	1	5	
243	0	3.3		**530**	1	6.6	
245	0	5.7	240	**532**	1	5.5	
247	0	1.2		**535**	1	6	
252	0	0.4		**537**	1	8.5	
257	0	5	185	**540**	1	4.4	
265	0	0.6		**542**	1	2.5	540
280	0	2.4		**545**	1	5	
282	0	3.3		**547**	1	3.4	
285	0	6	185	**550**	1	4.2	547
287	0	4		**552**	1	7.5	550
292	0	1.7		**555**	1	13.5	
297	0	2.4		**556**	1	6.2	550
300	0	1.2	297	**557**	1	1.5	556
301	0	1.2	300	**559**	1	4	
302	0	3.7		**560**	1	6.5	516
305	0	4		**562**	1	4.5	560

Table 10.2. Station's workload for different line configurations

(NbS_1, NbS_2)	Link	WkC_1	WkC_2
(1, 1)	no	329.6	200.7
	(1, 1)	265.4	264.9
(1, 2)	no	329.6	100, 100
	(1, 2)	176.8	178, 175.5
(2, 1)	no	164.8, 164.8	200
	(2, 1)	176.8, 176.7	176.8
(2, 2)	no	164.9, 164.7	100, 100
	(2, 2)	132.8, 132.4	141.9, 123.2
(2, 3)	no	194, 164	73, 68, 58
	(2, 3)	106, 106	94, 106, 117
(3, 1)	no	109.9, 109.9, 109.8	200.7
	(3, 1)	132.6, 132.6, 132.4	132.7
(3, 2)	no	109.9, 109.9, 109.8	100, 100
	(3, 2)	106.3, 106.3, 105.8	107.5, 104.6
(3, 3)	no	110.3, 109.8, 109.5	67.7, 53.3, 79.7
	(3, 3)	88, 88, 88	88, 89, 87
(4, 1)	no	82, 82, 82, 82	200
	(4, 1)	106, 106, 106, 105	106
(4, 2)	no	82, 82, 82, 82	100, 100
	(4, 2)	88, 87, 88, 88	92, 84
(4, 3)	no	82, 82, 82, 82	73, 68, 58
	(4, 3)	75, 76, 75, 75	76, 76, 74
(4, 4)	no	82.3, 82.4, 82.5, 82.4	56.8, 31.5, 68.7, 43.7
	(4, 4)	66.3, 66.3, 66.4, 68.9	65.2, 55.8, 75.2, 66.2

account the architecture of the AL can help to balance different workcentres. The idea is to analyse the flow of products between workcentres and to use the links between them to allow transferring some tasks. This helps to smooth the station's workload.

10.4.2 Resource Planning Application: Car Alternator

The chosen product is a car alternator, which corresponds to a real industrial case (Figure 10.7). The desired cycle time of the AL is fixed to 15 seconds.

Table 10.3. Stations workload (NbS_1=4, NbS_2=3)

no link		
WkC	PT	Ops
0	82	180, 182, 185, 347, 257, 60, 240, 197, 193, 310, 410, 147, 297, 120, 135, 62, 22, 106, 417, 170, 122, 40, 190, 35, 252, 285, 65, 241, 355, 245
0	82	415, 152, 300, 125, 155, 157, 70, 27, 225, 110, 130, 172, 430, 50, 162, 42, 312, 75, 112, 30, 132, 165, 87
0	82	47, 90, 127, 280, 247, 367, 237, 350, 243, 265, 292, 387, 282, 338, 357, 420, 425, 305, 52, 137, 320, 337, 97, 335, 12, 117, 45, 301, 17, 55, 102, 332, 325, 302, 327
0	82	427, 372, 380, 339, 397, 32, 67, 167, 307, 37, 20, 115, 15, 317, 287, 160, 322, 330, 100, 227, 140, 25, 395, 400, 402, 405
1	73	450, 452, 455, 517, 520, 522, 525, 527, 530, 532, 535, 537, 540, 460, 457
1	68	462, 465, 467, 470, 475, 477, 480, 482, 485, 487, 490, 507, 512, 515, 516
1	58	550, 556, 557, 559, 562, 552, 555, 560, 542, 545, 547

with link		
WkC	PT	Ops
0	75	180, 182, 185, 60, 240, 147, 135, 297, 417, 130, 152, 137, 40, 245, 197, 27, 32, 397, 132, 355, 310, 252, 50
0	76	193, 120, 62, 170, 190, 122, 162, 430, 155, 140, 115, 106, 227, 300, 287, 97, 327, 167, 320, 305, 65, 257, 285, 241, 325, 70
0	75	330, 110, 165, 225, 322, 37, 372, 337, 243, 47, 90, 280, 247, 350, 282, 100, 75, 301, 67, 335, 317, 347, 367, 117, 410, 22
0	75	338, 357, 415, 339, 332, 160, 17, 125, 52, 15, 20, 102, 172, 25, 302, 45, 307, 55, 265, 387, 30, 157, 112, 312, 380, 87, 12, 35, 556, 42
1	76	450, 452, 455, 517, 520, 522, 525, 527, 530, 532, 535, 537, 540, 542, 545, 547
1	76	457, 465, 467, 470, 475, 477, 480, 482, 485, 487, 490, 507, 512, 515, 516, 460, 462
1	74	550, 557, 559, 560, 562, 552, 555, 127, 420, 427, 425, 292, 237, 395, 400, 402, 405

A description of the operations performed on the product is summarised in Tables 10.5, 10.6 and 10.7. Table 10.4 presents the precedence constraints between tasks. Table 10.5 presents for each task the possible resources needed to accomplish this task and the operating mode (M: manual; R: robotic; A: automated) associated with each piece of equipment. For instance, task 1 can use one of the three pieces of equipment $\{0, 1, 2\}$, 0 being done manually, whilst 1 and 2 are automated FGs. Table 10.6 shows the process time and the cost of each piece of equipment associated with a given operation. The last column in the table shows for each piece of equipment the number of necessary operators (one operator for manual tasks and none in the case of machines or robots). The input data were prepared and structured using the SELEQ software package [110]. Only two criteria were optimised in this example:

- imbalance of workload – the imbalance between the process time of the stations has to be minimised;
- cost – the price of the AL has to be minimised.

Figure 10.7. A view of the car alternator

Table 10.4. Precedence constraints of the product

Op	Preds	Op	Preds	Op	Preds
1	4	**17**	16	**33**	32
2	1	**18**	6	**34**	31
3	2	**19**	17	**35**	33, 34
4	-	**20**	19	**36**	35
5	3	**21**	20	**37**	22
6	10	**22**	21	**38**	31
7	-	**23**	17	**39**	36, 38, 46, 47, 48
8	-	**24**	17	**40**	39
9	-	**25**	18	**41**	40
10	7, 8, 9	**26**	16	**42**	41
11	10	**27**	44	**43**	42
12	10	**28**	45	**44**	29, 30
13	11, 12	**29**	28	**45**	20, 23, 24, 25
14	13	**30**	28	**46**	35
15	14	**31**	27	**47**	35
16	15	**32**	31	**48**	35

Note that the real number of stations cannot be determined by computing the ratio between the sum of the operating times and the cycle time. Indeed,

Table 10.5. Operating mode and possible resources associated with each task

Op	MODE	EQUIP	Op	MODE	EQUIP
1	M	0	**24**	M	35
	A	1	**25**	A	36
	A	2		M	37
2	M	3	**26**	A	38
	A	4		A	39
	R	5	**27**	M	41
3	A	6	**28**	M	42
	M	7	**29**	A	43
4	M	8		A	44
5	M	9	**30**	A	45
6	M	10		A	46
7	M	11	**31**	M	47
8	M	12	**32**	M	48
9	M	13	**33**	A	49
10	A	14		A	50
11	A	15		A	51
12	A	16	**34**	A	52
13	R	17		A	53
	R	18	**35**	A	54
14	R	19	**36**	A	55
	R	20	**37**	M	56
15	A	21	**38**	A	57
	A	22	**39**	M	58
16	A	23	**40**	M	59
		24		A	60
17	R	25		A	61
	R	26		A	62
	R	27	**41**	A	63
18	M	28	**42**	M	65
19	M	29	**43**	M	66
20	A	30	**44**	M	67
21	A	31	**45**	A	68
	M	32	**46**	A	69
22	M	33	**47**	A	70
23	M	34	**48**	A	71

that number constitutes the theoretical minimum number of stations without considering the precedence constraints and the operating mode of the operations. The cycle time constraint is complied with by observing that there is a minimal/maximal duration for each task. The theoretical minimal (maximal) number of stations is the sum of the duration of the fastest (slowest) resource of each task over the cycle time. For the case presented here, the theoretical minimum number of stations is equal to 22, while the maximal number is 25.

In order to generate possible solutions, the ICA presented in Chapter 7 is used. The MOGGA was applied to this example for several user preferences. The results of the method are examined for different weight combinations corresponding to the relative importance one might give to each objective.

Table 10.6. Process time, cost (arbitrary units) and number of operators required by each piece of equipment

Equip	Time	Cost	NB_OP	Equip	Time	Cost	NB_OP
0	800	1712023	1	**35**	900	1700000	0
1	700	118396	0	**36**	400	80687	0
2	800	131218	0	**37**	500	1835082	0
3	400	1700000	1	**38**	1500	99613	0
4	200	100484	0	**39**	1400	476287	0
5	200	344492	0	**40**	900	1775000	1
6	400	466587	0	**41**	1500	1700000	1
7	1500	1795355	1	**42**	600	92387	0
8	300	1700000	1	**43**	700	468292	0
9	300	1700000	1	**44**	800	90403	0
10	300	1700000	1	**45**	800	468292	0
11	300	1700000	1	**46**	300	1700000	1
12	300	1700000	1	**47**	300	1700000	1
13	300	1700000	1	**48**	600	114550	0
14	1500	125000	0	**49**	600	488751	0
15	0	83931	0	**50**	700	198341	0
16	0	83931	0	**51**	400	10424	0
17	600	35915	0	**52**	400	45570	0
18	700	328029	0	**53**	1500	75000	0
19	600	18926	0	**54**	300	70000	0
20	700	471996	0	**55**	300	1700000	1
21	200	6473	0	**56**	1400	75000	0
22	300	384361	0	**57**	300	1700000	1
23	800	77318	0	**58**	1000	1700000	1
24	900	231324	0	**59**	500	79298	0
25	500	27659	0	**60**	500	81960	0
26	700	271667	0	**61**	600	457187	0
27	600	172932	0	**62**	1400	25000	0
28	400	1700000	0	**63**	1500	1700000	1
29	800	1700000	0	**64**	1500	1700000	1
30	800	45570	0	**65**	300	1700000	1
31	400	80687	0	**66**	1400	37500	0
32	500	1835082	0	**67**	400	70000	0
33	500	1700000	0	**68**	400	70000	0
34	500	1700000	0	**69**	400	70000	0

Table 10.7 summarises the results obtained. It presents the process times on the different stations and the total cost of the line according to the different

optimisation strategies. The number of stations is given by N, the cost of the line by 'Cost' and the balancing by 'Bal.'. The columns labelled from 1 to 25 represent the workload of the different stations. Bold numbers represent stations where the cycle time is exceeded. The weight attributed to the balancing is B and the one for cost being C. The weights (B, C) represent the relative importance given to each criterion. In this case, three pairs of preferences, *i.e.* $\{(0,1),(1,0),(0.5,0.5)\}$, were used. The pair $(0,1)$ means that the cost is the only important objective; no care is given to the imbalance of the line. In contrast, the pair $(1,0)$ means the opposite. Finally, the pair $(0.5,0.5)$ means that the same importance is given to the two objectives.

Table 10.7. Process time of each station according to the different weights (B, C)

N	B	C	1	2	3	4	5	6	7	8	9	10	11	12	13	14	15	16	17	18	19	20	21	22	23	24	25	Bal.	Cost
22	0	1	**23**	22	15	12	15	13	13	12	14	15	14	18	14	**5**	15	15	14	16	15	14	15	15				254	22148352
22	0.5	0.5	15	15	15	15	14	13	14	**12**	12	14	15	14	18	14	15	15	14	**21**	15	14	15	15				74	23848352
22	1	0	15	15	15	15	15	15	14	**12**	13	14	15	15	18	14	15	15	15	**21**	15	14	15	15				62	29197448
23	0	1	**23**	22	15	14	13	4	**4**	8	12	14	14	15	14	23	10	15	15	14	16	15	14	15	15			513	22148352
23	0.5	0.5	10	14	15	12	15	14	13	13	12	14	14	15	14	**18**	14	**10**	15	15	16	15	14	15	15			93	24068032
23	1	0	15	15	15	15	15	15	12	**16**	14	14	15	15	12	15	14	**11**	15	15	14	13	14	15	15			44	28657204
24	0	1	15	9	**18**	15	14	13	14	12	12	14	15	12	6	18	10	15	15	14	15	11	**5**	14	15	15		312	22437168
24	0.5	0.5	10	14	15	12	15	14	13	**9**	13	12	14	**15**	14	12	14	10	14	15	15	15	13	14	15	15		132	25768032
24	1	0	15	15	**15**	15	14	15	14	12	13	14	15	13	**7**	15	11	11	15	15	14	13	15	14	15	15		122	30675056
25	0	1	12	9	**21**	15	4	14	13	9	13	12	14	15	14	18	**4**	5	14	10	15	15	16	15	14	15	15	516	22355208
25	0.5	0.5	14	7	15	9	**15**	14	13	13	13	12	14	15	14	12	**4**	14	10	14	15	15	13	15	14	15	15	287	27248352
25	1	0	14	**8**	15	12	15	14	12	12	9	13	13	14	15	15	11	15	11	**15**	15	14	13	15	14	15	15	161	33150960

The algorithm was run 12 times using four different 'number of stations' N (N varying from the theoretical minimum number of stations to the theoretical maximal number) and three combinations of preferences. The results show that the proposed method satisfies the user's preferences with regard to the optimisation objective. Figure 10.8 shows the cost of the line according to the number of stations for several preferences. This demonstrates that the increase of the cost with the number of stations is not a general behaviour. For instance, the cost of a line with 23 stations is less than with 22 stations (for weights set to $(1,0)$). For a given number of stations, the cost of the line corresponding to $(1,0)$ is high in comparison with $(0.5,0.5)$, which is higher than the cost corresponding to $(0,1)$.

The results corresponding to the solutions with 24 stations allow one to make the following comments:

- The couple $(B=1, C=0)$ yields a minimal process time of 7 (station 13), a maximal process time of 15 and a cost of 30675056.
- The couple $(B=0.5, C=0.5)$ yields a minimal process time of 9 (station 8), a maximal process time of 15 and a cost of 25768032.
- The couple $(B=0, C=1)$ yields a minimal process time of 5 (station 21), a maximal process time of 18 (station 3) and a cost of 22437168.

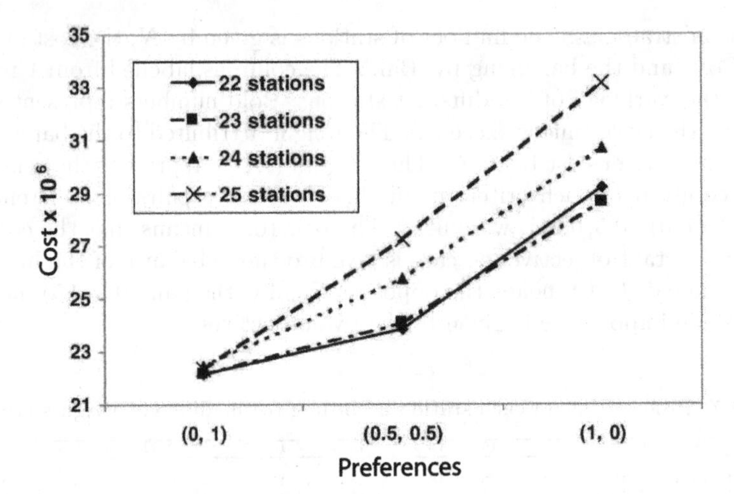

Figure 10.8. Cost (arbitrary units) of the line according to three preferences

The preference $(1, 0)$ yields an expensive but well-balanced line in comparison with other preferences (see Table 10.7). In contrast, the results obtained using the preference $(0.5, 0.5)$ show clearly that setting an equal weight to the two objectives does not mean that one will obtain the line with the lowest cost and the lowest imbalance simultaneously, but rather the best compromise between the two objectives. Finally, the pair $(0, 1)$ leads to a cheapest (minimal cost) and a less-balanced line. Figure 10.9 shows that the preference $(1, 0)$ leads to a good balancing in comparison to the other ones.

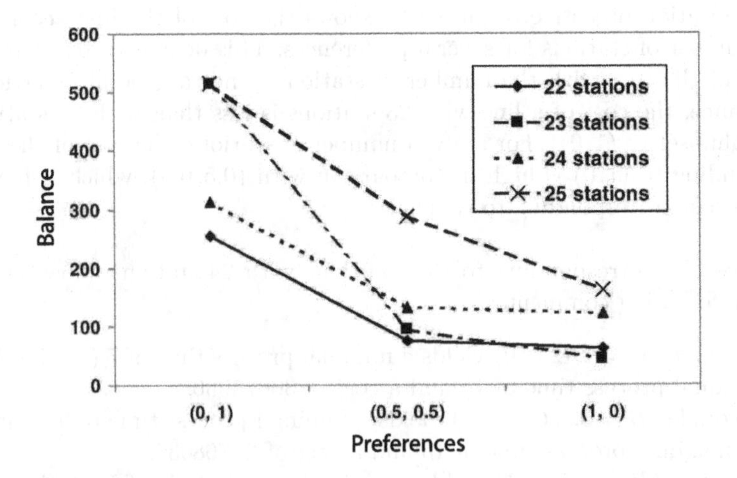

Figure 10.9. Balancing of the line according to the preferences set for different number of stations

The station load can exceed the cycle time in some cases. This means that the desired cycle time cannot be held for the selected number of stations. The line will generally be less expensive as the cycle time constraint is relaxed. These results show that a solution using 22 stations leads to the cheapest cost if the balancing is not important. The choice of a solution is user dependent. A good compromise between the balancing and the cost of the line corresponds to a solution with 24 stations using the $(0.5, 0.5)$ preference.

The analysis of the three combinations of weights clearly demonstrated that considering each criterion separately leads to very bad results according to the other ones. Given the same preference for the two objectives leads to solutions where the values obtained for each criterion lead to good compromises between them.

The main advantage of such a computer-aided tool is that it allows one to try a lot of different combinations for a lot of different sets of data. This is almost impossible to realise manually due to the very large amount of possible solutions. An important aspect of this approach is that the DM controls the optimisation process.

11

A Real-world Example Optimised by the OptiLine Software

The concepts presented in this book are illustrated in the following case study, which is supplied by a major European car manufacturer. The production involved two different cars (Model1 and Model2), each with various options, making up 10 different car models, with the percentages of occurrence in the total production being as shown in Figure 11.1

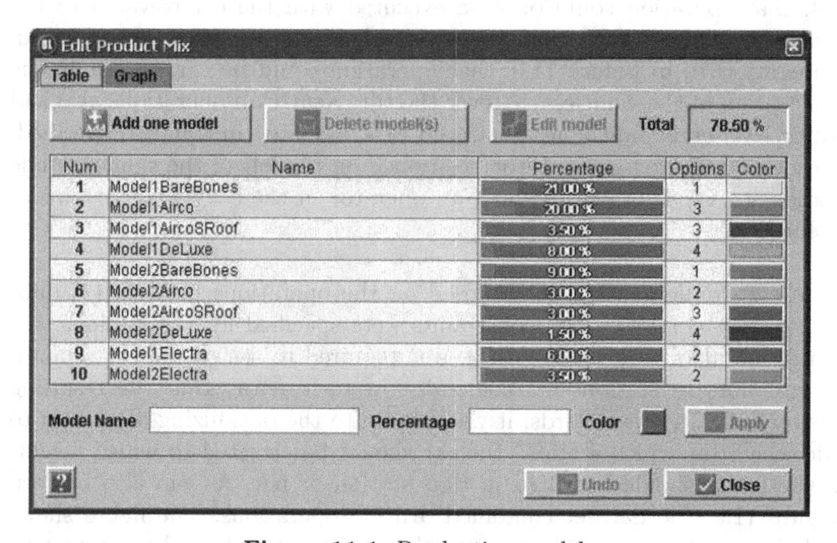

Figure 11.1. Production models

As can be seen in the third column (percentage) of Figure 11.1, some of the models make up a very small part of the whole production, giving rise to the difficult phenomenon of 'rare models'; clearly, line balancing 'on average' would not do, and care will have to be taken of peak times.

The precedence graph of the whole production is depicted in Figure 11.2, with operations pertaining to specific options in different shades. The two cars are in the upper and lower parts of the graph respectively. Many operations are common to all models in the same car, while others pertain to the various options. The similarity of the graphs for the two cars is no coincidence, as the structure of the cars is basically identical, while involving mounting different parts for each of the cars, yielding operations of different durations. There is, nevertheless, a small subset of the operations that are carried out whatever car or model is being assembled – these are in the middle-right of the figure in grey. As pointed out earlier, the automotive industry makes extensive use of ergonomic constraints, and this dataset was no exception to that rule. All operations had some ergonomic constraint imposed on them, as can be seen in the far right column of the operation list in Figure 11.3.

For most of the operations, two station-level ergonomic constraints (WLECs, depicted in circles) were specified: the elevation of the vehicle and its tilt, as was the case for instance for operation 73 in the figure (first line of the table); that operation could only be executed with the car elevated to the medium height, and tilted. Some operations had only the elevation specified, *e.g.* operation 76 in Figure 11.3; that operation could be carried out in the high elevation of the vehicle, whatever the tilt. Nearly all operations also had operator-level ergonomic constraints (OLECs, depicted in squares) specified: operation 73 required an operator positioned on the left of the vehicle, while operation 91, for example, required an operator in the centre (*i.e.* cabin) of the vehicle.

The ergonomic constraints imposed on the operations presented a difficulty, since no corresponding constraints were specified for the stations: the optimization algorithm in OptiLine was required to assign the corresponding constraints to stations automatically, in such a way that the resulting line is optimal. In other words, it was left up to the optimizing algorithm to decide where (at which station) the car should be elevated to which height, and whether it should be tilted in that station or not. As can also be seen in Figure 11.3, the dataset contained drifting operations. The figure shows two of them (76 through 78, and 106 plus 107), but in fact there were four of them, two for each of the cars. Since these operations (namely mounting the air conditioning and the sunroof respectively) required specific equipment, they were linked together, thus ensuring they would be carried out (the airco, and the sunroof) on the same stations. However, the actual identity of the stations carrying them out was once again not imposed; this was left to the algorithm to find out for itself, so the resulting line would be optimal.

Some operations were subject to zoning constraints, *e.g.* as was the case for operations 93 and 94 in Figure 11.3; these operations could only be placed on a specified subset of the station. For operations 93 and 94, that subset was

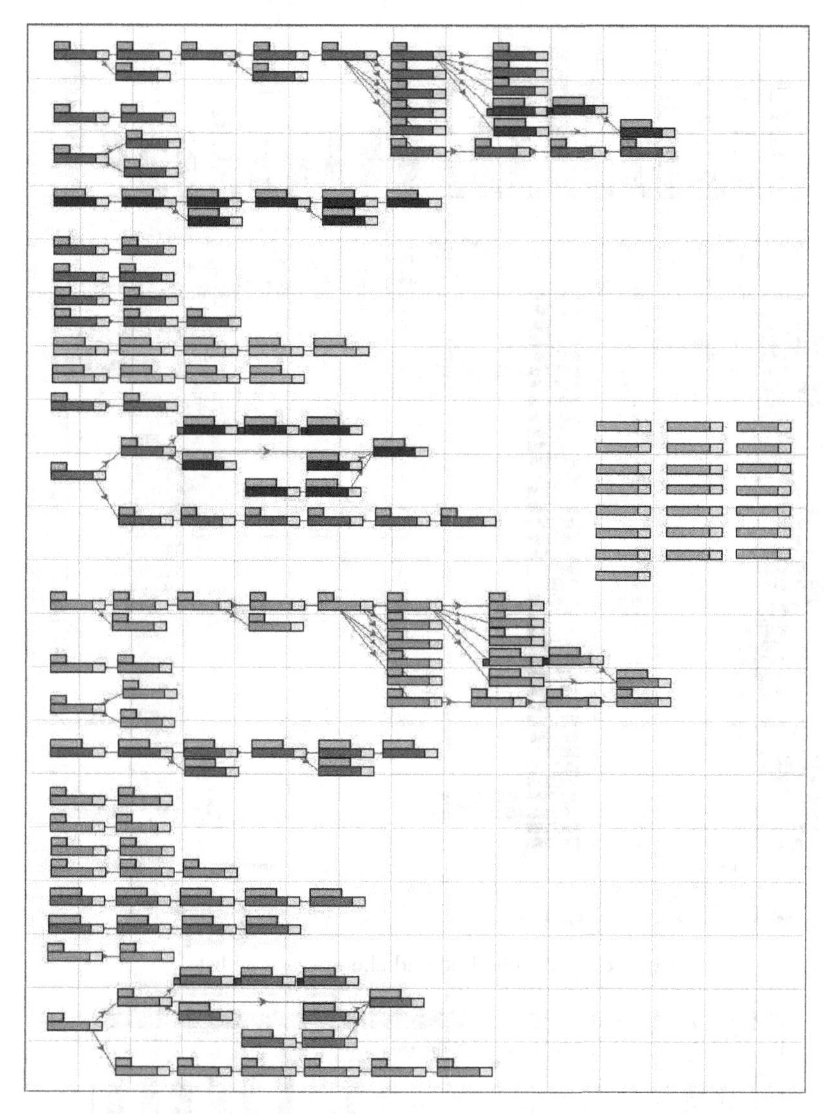

Figure 11.2. The precedence graph

the 'end' of the line, as depicted in Figure 11.4; these operations were allowed to be placed by the algorithm only onto stations 12 through 20.

Some operations required the simultaneous cooperation of two operators, *e.g.* operations 100 and 101 in Figure 11.3. In nearly all cases the positions of the operators were also imposed on the operation (left and right of the car for operations 100 and 101), as illustrated for the first of the two operators in

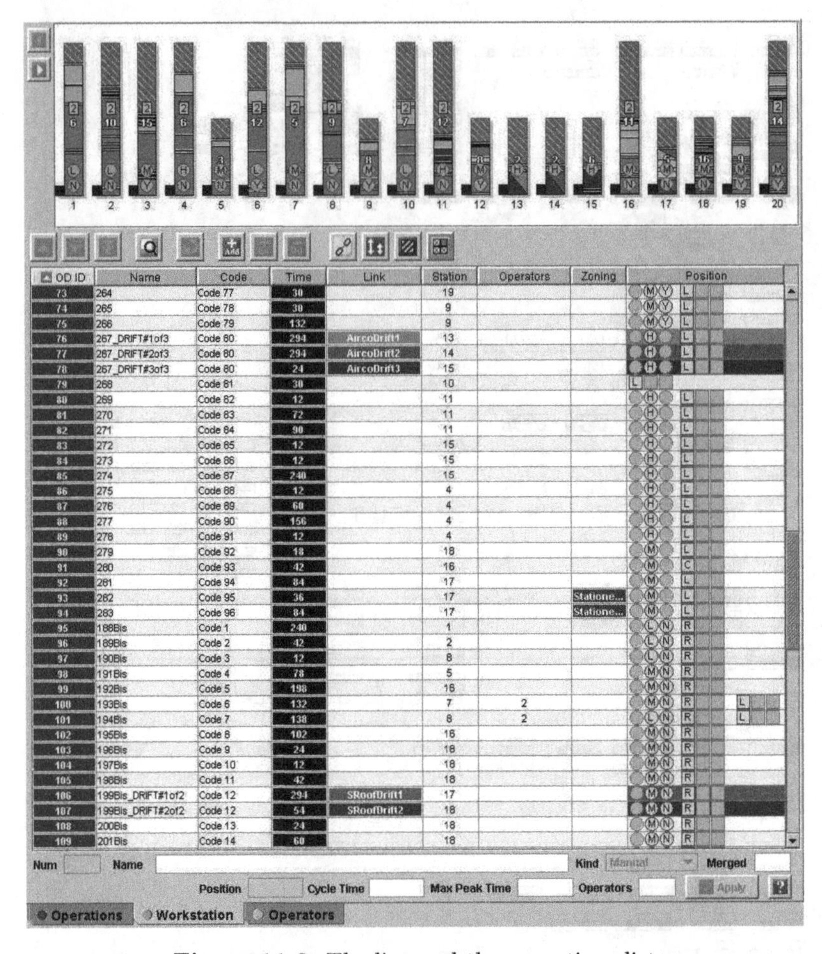

Figure 11.3. The line and the operations list

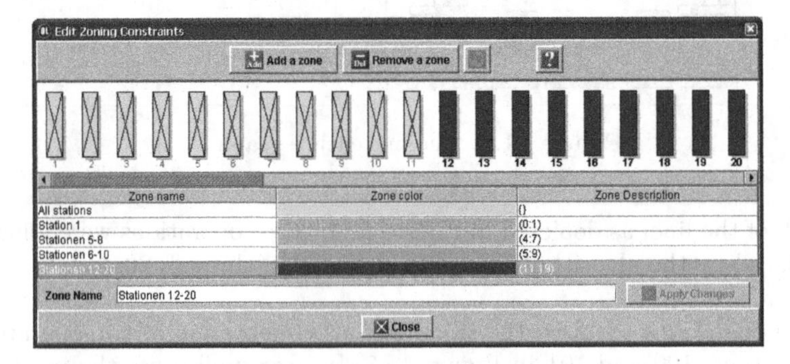

Figure 11.4. Zoning constraints

Figure 11.5. This meant that the algorithm was required to allocate at least those two operators in the station that would carry out those operations.

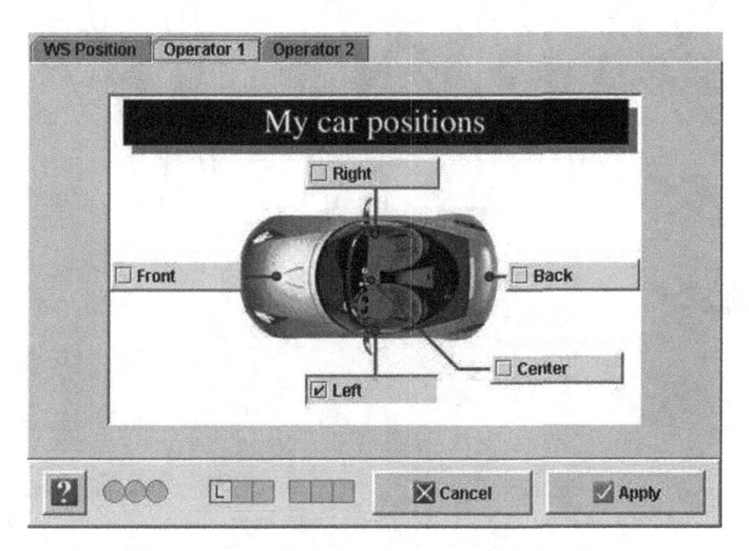

Figure 11.5. Specifying ergonomic constraints

The presence of the numerous constraints made this a very difficult dataset to solve. Nevertheless, OPTILINE's optimizing algorithm supplied in a few minutes a solution compliant with all of them; this is depicted in the top panel of Figure 11.3. Note the circles in the stations: their shade indicates that the particular ergonomic constraints were not specified *a priori* by the user, but supplied automatically by the algorithm.

That panel shows the average behaviour of the line, confirming that all stations are, on average, well within the cycle time (the spare time is depicted by a green segment in the top of the station bar). Some of the stations were given two operators, which is depicted in Figure 11.3 by their bars having double the height of the single-operator stations.

Although it is reassuring to have the average behaviour of the line within the cycle time, it is not enough for a well-managed line, as discussed earlier. It is therefore reassuring that the solution supplied by OPTILINE also took care of peak times. This is illustrated in Figure 11.6, where the time taken by each model assembled on the line is computed for each station.

As can be seen in the figure, the solution supplied by OPTILINE kept the total station time for any model being assembled within the cycle time. Consequently, no stoppage should occur if the line is fed the product mix specified in Figure 11.1, provided the models are scheduled to accommodate the four

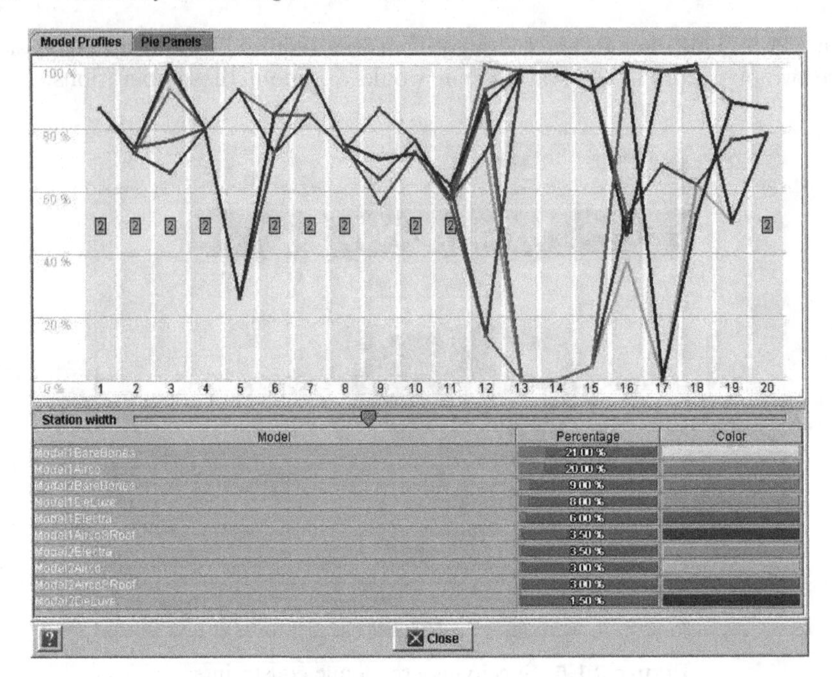

Figure 11.6. The transitory behaviour (peak times)

drifting operations discussed above. Needless to say, that result took into account the fact that, in stations having more than one operator, the execution time is the result of a fully fledged within-station scheduling, rather that a simple sum of operation durations. In addition, that scheduling must comply with any ergonomic constraints imposed on the operation, as well as precedence constraints.

An example of such a scheduling is given in Figure 11.7, showing the scheduling for one model inside a station featuring two operators. The arrow indicates a precedence constraint among the operations. The figure depicts the operations that will be carried out by the station each time a car model ‘Model2DeLuxe’ is assembled by that station. The different shades of the operations in the bars relate to the different options of the car features. The operations barred in the bottom table are the operations that are also carried out in that station, but do not pertain to the particular ‘Model2DeLuxe’ being shown.

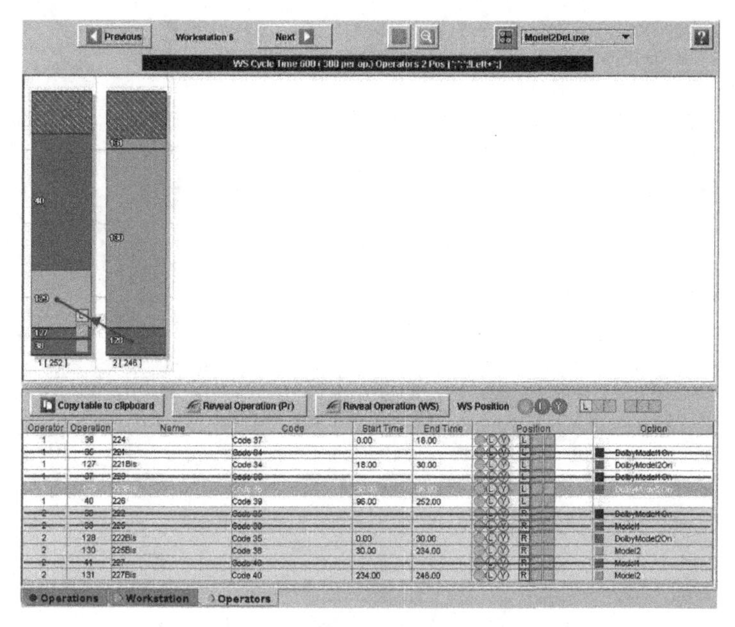

Figure 11.7. Within-station scheduling

12

Conclusions and Future Work

12.1 We Attained...

In order to deal with ALB, a new algorithm called EPAL is introduced. The hard constraint of the problem is the fixed number of stations and the aim is to find the best balanced assembly system. The proposed approach is based on the so-called 'boundary stones'. Also, several heuristics are embedded in a GGA. In order to deal with the changes during the operation phase of the AL, a new concept of BFO is introduced. This concept allows the user to treat the balancing and the scheduling model at the design phase. In the case of an HAL, the RP aims to select equipment to carry out the assembly tasks. A new method is presented which is based on the MOGGA and PROMETHEE II method. The focus is on how to deal with a user's preferences in design problems.

12.2 Tendencies and Orientations

Billions of dollars have been spent annually on the construction of new facilities. It is estimated that the layout of most facilities is modified approximately every 2 to 5 years. This continuous change is required to keep pace with changes in demand, new product introductions, process changes, improved tooling and technology, new legislation, *etc.* With the increased diversified demand in production, manufacturers tend, depending on the type of production, to use mixed-model ALs or batch production. The evolution of the demand forces designers to reconfigure their ALs. Techniques that allow one to deal with the evolution of the architecture of ALs over the time are required more than the methods that propose designs from *scratch*. This involves an increasing need for fast computer-aided design tools to follow the frequent changes. The aim is to develop ALD tools which allow one to consider this evolution. In order to design ALs efficiently, knowledge about task complexity, preferences on task grouping, and the process time must be taken into account. An interactive and iterative method '*design for humans*' can allow the introduction of such knowledge to computer-aided design methods. The designers propose a set of alternatives of the AL, while operators give their experience and criticism on the proposed solutions.

12.3 Data Collection

Most of the industrial approaches for design problems suffer from the amount of data to be used. On the other hand, existing academic algorithms require little input data and cannot be applied to industrial problems. Thus, there is a great demand to combine both approaches. An intensive collaboration with industry is required to collect the data needed to model the full real-world AL design problem.

12.4 Model Formulation

The line layout module assigns tasks to stations and decides about the position of stations and resources on the plant floor. The LL assigns tasks to stations, while the PL module determines the space requirements taking into account station dimensions and material storage, handling systems, *etc.* The line layout problem is an iterative and interactive procedure; this is illustrated in Figure 12.1 and can be described as follows:

- The production approach helps to decide about batch and mixed production.
- The line description is the input data of the design method. It allows one to describe the problem, preferences, and constraints.
- The design objectives allow one to express the criteria to optimise, as well as designer's desiderata.
- The design tool proposes the AL architecture taking into account designer preferences.
- The simulation module permits the verification of the proposed design.

12.5 Validation and Output Analysis

Performance evaluation generally involves two steps: (1) the mathematical model and (2) the model solution. Since it is difficult to find a simple model to describe a studied system, a simulation method must be used. A standardisation of *performance indices* of line layout design must be defined, as well as the factors that may affect the performance of the system.

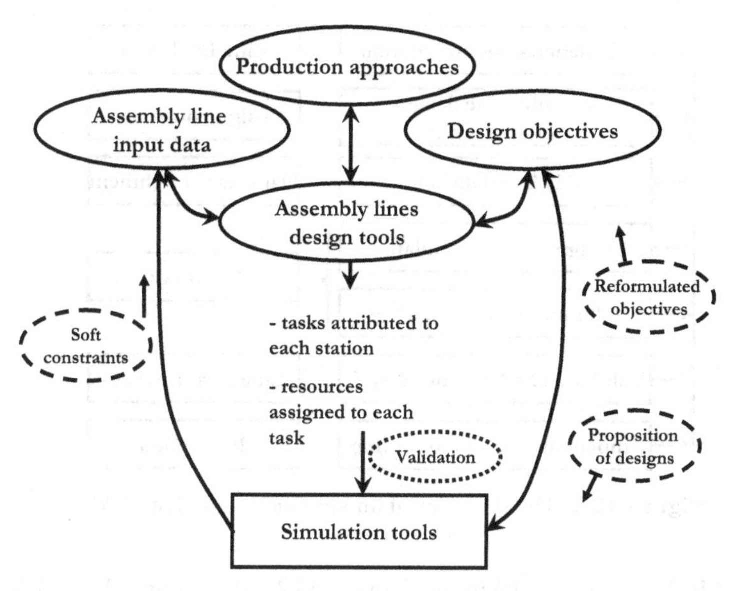

Figure 12.1. An integrated method to AL design

12.6 The Proposed Approach

A project composed of seven phases with the following structure is proposed (Figure 12.2) as follows.

Line Evolution. The objectives of this task are the identification and the modelling of the evolution of the AL. Line designers and resource planners can help to understand the phenomenon. This phase will reduce the gap between design industries' approaches and academic ones.

Design Constraints. The aim of this task is the interaction between designers and workers. Human factors (like task complexity, reliability, worker's experience, human skills) must be studied and integrated into the design process. Depending on the production approaches, objectives, and constraints, the complexity to find a good solution may vary.

Database Enrichment. The aim of this task is to describe efficiently the line design problem. An enrichment of a database with information to help design tools is required. Useful information must be stored in database, while useless information must be discarded. Also, research in the field of graphical representation of graphs for complex products has to be done. The aim is to help designers to get an idea quickly of the implication of these graphs on the possible balancing.

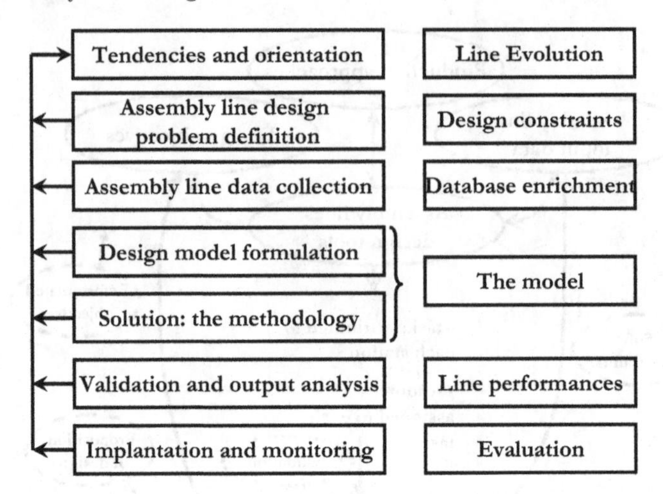

Figure 12.2. Development of an approach to design of ALs

The Model. Once all the input data of the AL design have been collected and verified, the next step is to model the design tool: the output data, the interaction between the different modules, the methods to develop, *etc.* The model has to consider the results of the tasks cited above.

Line Performances. The aim of this task is to define the performance indices of line layout design. Since ALD is a multiple objective search problem, the goal is to define an easy way to select the best line design from existing ones. The method will be based on MCDA methods. The aim is to exploit the features of GAs to allow designers to deal effectively with user preferences.

Evaluation. A user-friendly interface must be developed in order to facilitate the access to the AL data stored in many different databases. In order to validate the different algorithms and methods, design tools have to be integrated in these design packages, such as CATIA V5, and simulation models, such as AUTOMOD.

References

1. A. Agnetis and C. Arbib. 'concurrent operations assignment and sequencing of particular assembly problems in flow lines'. *Annals of Operations Research*, 69:1–31, 1997.
2. A. Agnetis, A. Ciacimino, M. Pizzichella, and M. Lucertini. 'task synchronization in flexible system for car components assembly'. *In 'Optimization in industry 2', Edited by Ciriani T.A. and Leachman R.C., John Wiley & Sons Ltd*, 1994.
3. A. Agnetis, F. Nicolò, C. Arbib, and M. Lucertini. 'task assignment and subassembly scheduling in flexible assembly lines'. *IEEE Transactions on Robotics and Automation*, 11(1):1–20, 1995.
4. E. J. Anderson and M. C. Ferris. 'genetic algorithms for combinatorial optimisation: the assembly line balancing problem'. *ORSA Journal on Computing*, 6:161–173, 1994.
5. C. Arbib, G. Ciaschetti, and F. Rossi. 'distributing material flows in a manufacturing system with large product mix: two models based on column generation'. *Lecture Notes in Economics and Mathematical Systems, Springer-Verlag*, 480:235–249, 1999.
6. A. L. Arcus. 'comsoal: a computer method of sequencing operations for assembly lines'. *International Journal of Production Research*, 4:259–277, 1966.
7. R. G. Askin and C. R. Standridge. 'modelling and analysis of manufacturing systems'. *John Wiley & Sons Inc., New York*, 1993.
8. J. Baker. 'reducing bias and inefficiency in the selection algorithm'. *Proceedings of the second international conference on genetic algorithms, San Mateo, July*, 1987.
9. J. F. Bard. 'assembly line balancing with parallel stations and dead time'. *Int. J. Prod. Res.*, 27(6):1005–1018, 1989.
10. J. F. Bard, E. Dar-El, and A. Shtub. 'an analytic framework for sequencing mixed model assembly lines'. *Int. J. of Prod. Res.*, 30(1):35–48, 1992.
11. J. J. Bartholdi and Eisenstein D. D. 'a production line that balances itself'. *Operations Research*, 44(1):21–34, 1996.
12. I. Baybars. 'a survey of exact algorithms for the simple assembly line balancing'. *Management Science*, 32:909–932, 1986.

13. T. Bäck. 'evolutionary algorithms in theory and practice: evolution strategies, evolution programming, genetic algorithms'. *Oxford University Press, New York*, 1996.
14. P. J. Bentley. *'Generic evolutionary design of solid objects using a genetic algorithm'*. PhD thesis, Division of Computing and Control Systems, School of Engineering, University of Huddersfield, 1996.
15. I. Berger, J-M. Bourjoully, and G. Laporte. 'branch and bound algorithms for the multi-product assembly line balancing problem'. *European Journal of Operational Research*, 58:215–222, 1992.
16. S. Bock and O. Rosenberg. 'a new distributed fault-tolerant algorithm for the simple assembly line balancing problem 1'. *Technical report TR-RSFB-97-040, University of Paderborn*, 1997.
17. F. F. Boctor. 'a multiple-rule heuristic for assembly line balancing'. *J. of Oper. Res. Society*, 46(1):62–69, 1995.
18. G. Boothroyd. 'assembly automation and product design'. *Marcel Dekker Inc., New York*, 1992.
19. E. H. Bowman. 'assembly line balancing by linear programming'. *Oper. Res.*, 8(3):385–389, 1960.
20. J.-P. Brans and B. Mareschal. 'the promcalc & gaia decision support system for multicriteria decision aid'. *Decision Support Systems*, 12:297–310, 1994.
21. J. Bukchin. 'a comparative study of performance measures for throughput of a mixed model assembly line in a JIT environment'. *Int. J. of Prod. Res.*, 36(10):2269–2685, 1998.
22. J. Bukchin and M. Tzur. 'design of flexible assembly line to minimise equipment cost'. *IIE Transactions*, 32(7):585–598, 2000.
23. J. W. Burrow. 'darwin the origin of species'. *Pelican classics, Middlesex, England*, 1979.
24. B. J. Carnahan, M. S. Redfen, and B. A. Norman. 'incorporating physical demand criteria into assembly line balancing'. *Internal Report TR99-8, Pittsburgh University, Department of Industrial Engineering*, 1999.
25. P. Chevalier, B. Raucent, and P. Semal. 'optimizing the design of an assembly line'. *Proceeding of FAIM'99, Tilburg, Netherlands*, pages 415–424, 1999.
26. W. Choi and Y. Lee. 'line balancing of mixed-model assembly lines. *Proceedings of EDA'98 (CD-ROM), Hawaii*, 1998.
27. W.-M. Chow. 'assembly line design methodology and applications'. *Marcel Dekker Inc., New York*, 1990.
28. P. C. H. Chu. *'A genetic algorithm approach to combinatorial optimisation problems'.* PhD thesis, The Management School, Imperial College of Science, Technology and Medicine, London, England, 1997.
29. C. A. C. Coello. 'a comprehensive survey of evolutionary-based multiobjective optimization'. *Knowledge and Information Systems*, 1(3):129–156, 1999.
30. J. C. Culberson. 'on the futility of blind search: an algorithmic view of "no free lunch"'. *Evolutionary Computation*, 6(2):109–127, 1998.
31. D. Cvetković and I. C. Parmee. 'use of preferences for GA-based multi-objective optimisation'. *Proceedings of GECCO'99, Orlando USA*, pages 1504–1509, 1999.
32. L. Davis. 'applying adaptive algorithms to domains'. *Proceedings of the International Joint Conference on Artificial Intelligence*, pages 162–164, 1985.
33. L. Davis. 'handbook of genetic algorithms'. *Van Nostrand Reinhold, New York*, 1991.

34. A. K. De Jong. *'An analysis of the behavior of a class of genetic adaptive systems'*. PhD thesis, University of Michigan, 1975.
35. P. De Lit and A. Delchambre. 'integrated design of a product family and its assembly system'. *Kluwer Academic Publishers, Norwell, Massachusetts, First Edition*, 2003.
36. P. De Lit, B. Rekiek, F. Pellichero, A. Delchambre, J. Danloy, F. Petit, A. Leroy, J.-F. Marée, A. Spineux, and B. Raucent. 'a new philosophy of design of a product and its assembly line'. *Proceedings of ISATP'99, Porto, Portugal*, pages 381–386, 1999.
37. E. B. Dean and R. Unal. 'elements of designing for cost'. *Presented at The AIAA Aerospace Design Conference, Irvine CA, AIAA-92-1057*, 1992.
38. K. Deb. 'multi-objective genetic algorithms: problem difficulties and construction of test problems'. *Evolutionary Computation*, 7(3):205–230, 1999.
39. A. Delchambre. 'CAD method for industrial assembly: concurrent design of products, equipment and control systems'. *John Wiley & Sons Inc., Chichester, England*, 1996.
40. J. Driscoll and A. Abdel-Shaffi. 'a simulation approach to evaluating assembly line balancing solutions'. *Int. J. of Prod. Res.*, 23(5):975–985, 1985.
41. B. H. Faaland, T. D. Klastorin, T. G. Schmitt, and A. Shtub. 'assembly line balancing with resource dependent task times'. *Decision Sciences*, 23:343–364, 1992.
42. E. Falkenauer. 'solving equal piles with a grouping genetic algorithm'. *Eshelman L.J. (Ed.), Proceedings of the Sixth International Conference on Genetic Algorithms (ICGA95), Morgan Kaufmann Publishers, San Francisco*, pages 492–497, 1995.
43. E. Falkenauer. 'a hybrid grouping genetic algorithm for bin packing'. *Journal of Heuristics*, 2(1):5–30, 1996.
44. E. Falkenauer. 'genetic algorithms and grouping problems'. *John Wiley & Sons Inc., Chichester, First Edition*, 1998.
45. E. Falkenauer. 'applying evolutionary algorithms to real-world problems'. *In L.D. Davis, K. De Jong, M. Vose and L.D. Whitley (Eds.) Evolutionary Algorithms, IMA Volumes in Mathematics and its Applications, Springer Verlag*, 111, 1999.
46. E. Falkenauer and A. Delchambre. 'a genetic algorithm for bin packing and line balancing'. *Proceedings of the IEEE International Conference on Robotics and Automation (RA92), IEEE Computer Society Press*, pages 1186–1192, 1992.
47. C. J. L. Fernandez and M. P. Groover. 'mixed-model assembly line balancing and sequencing: a survey engineering'. *Design and Automation*, 1(1):33–42, 1995.
48. L. J. Fogel, A. J. Owens, and M. J. Walsh. 'artificial intelligence through simulated evolution'. *New York: Wiley*, 1966.
49. C. M. Fonseca and P. J. Fleming. 'an overview of evolutionary algorithms in multiobjective optimization'. *Evolutionary Computation*, 3(1):1–16, 1995.
50. M. P. Fourman. 'compaction of symbolic layout using genetic algorithms'. *In Genetic Algorithms and their Applications: Proceedings of the First International Conference on Genetic Algorithms, Lawrence Erlbaum*, pages 141–153, 1985.
51. R. L. Francis, L. F. McGinnis, and J. A. White. 'facilities layout and location: an analytical approach'. *Prentice Hall International Series in Industrial and Systems Engineering*, 1996.

52. N. Gaither. 'production and operations management'. *Belmon, CA: Duxbury*, 1996.
53. M. R. Garey and D. S. Johnson. 'computers and intractability – a guide to the theory of NP completeness'. *Freeman, San Francisco USA*, 1979.
54. M. Gen and R. Cheng. 'genetic algorithms & engineering design'. *John Wiley & Sons Inc, First Edition, Canada*, 1997.
55. S. Ghosh and R.J. Gagnon. 'a comprehensive literature review and analysis of the design, balancing and scheduling of assembly systems'. *Int. J. of Prod. Res.*, 27:637–670, 1989.
56. F. Glover and M. Laguna. 'tabu search'. *Kluwer Academic Publishers, Boston*, 1997.
57. D. E. Goldberg. 'genetic algorithms in search, optimization and machine learning'. *AddisonWesley Publishing Company Inc.*, 1989.
58. S. C. Graves and R. C. Holmes. 'equipment selection and task assignment for multiproduct assembly system design'. *International Journal of Flexible Manufacturing Systems*, 1:31–50, 1988.
59. S. C. Graves and B. W. Lamar. 'an integer programming procedure for assembly system design problems'. *Operations Research*, 31(3):522–545, 1983.
60. S. C. Graves and D. E. Withney. 'a mathematical programming procedure for equipment selection and system evaluation in programmable assembly'. *Proceedings of the IEEE Decision and Control*, pages 531–536, 1979.
61. R. E. Gustavson. 'design of cost-effective assembly systems'. *C.S. Draper Laboratory Report, N. P-2661, Cambridge*, 1986.
62. A. L. Gutjahr and N. K. Nemhauser. 'an algorithm for the line balancing problem'. *Management Science*, 11(2):308–315, 1964.
63. P. Hajela and C. Y. Lin. 'genetic search strategies in multicriterion optimal design'. *Structural Optimisation*, 4:99–107, 1992.
64. D. W. He and A. Kusiak. 'design of assembly systems for modular products'. *IEEE Transactions on Robotic and Automation*, 13(5):646–655, 1997.
65. W. B. Helgeson and D. P. Birnie. 'assembly line balancing using the ranked positional weight technique'. *J. Ind. Engng*, 12(6):394–398, 1961.
66. S. S. Heragu. 'group technology and cellular manufacturing'. *IEEE Transactions on Systems, Man, and Cybernetics*, 24(2), 1994.
67. T. R. Hoffmann. 'assembly line balancing with precedence constraints'. *Management Science*, 9:551–562, 1963.
68. T. R. Hoffmann. 'eureka: A hybrid system for assembly line balancing'. *Management Science*, 38:39–47, 1992.
69. J. H. Holland. 'adaptation in natural and artificial systems'. *University of Michigan Press, Ann Arbor*, 1975.
70. J. Horn and N. Nafpliotis. 'multiobjective optimisation using the niched Pareto genetic algorithm'. *Technical Report, IlliGAL Report 93005, University of Illinois, Urbana, Illinois, USA*, 1993.
71. C. J. Hyun, Y. Kim, and Y. K. Kim. 'a genetic algorithm for multiple objective sequencing problems in mixed assembly lines'. *Computers and Operations Research*, 25(7-8):675–690, 1998.
72. J. R. Jackson. 'a computing procedure for the line balancing problem'. *Management Science*, 2:261–271, 1956.
73. A. Jaszkiewicz. 'genetic local search for multiple objective combinatorial optimization'. *Research report, Institute of Computing Science, Pozna? University of Technology, RA-014/98*, 1998.

74. R. V. Johnson. 'optimally balancing large assembly lines with "fable"'. *Management Science*, 34:240–253, 1988.
75. S. Johnson. 'optimal two- and three-stage production schedules with setup times included'. *Naval Research Logistics Quarterly*, 1:61–68, 1954.
76. D. R. Jones and M. A. Beltramo. 'solving partitioning problems with genetic algorithms'. *Proceedings of the 4th International Conference on Genetic Algorithms, Morgan Kaufmann*, pages 442–449, 1991.
77. A. K. Kamrani, H. R. Parsaei, and D. H. Liles. 'planning, design, and analysis of cellular manufacturing systems'. *Elsevier Science B.V., Holland*, 1995.
78. M. D. Kilbridge and L. Wester. 'a heuristic method for assembly line balancing'. *Journal of Industrial Engineering*, 12:292–298, 1961.
79. Y. K. Kim, Y. J. Kim, and Y. Kim. 'genetic algorithms for assembly line balancing with various objectives'. *Computers & Industrial Engineering*, 30(3):397–409, 1996.
80. S. Kirkpatrick, C. D. Jr. Gelatt, and M. P. Vecchi. 'optimisation by simulated annealing'. *Science*, 220:671–680, 1983.
81. J. Kirton and E. Brooks. 'cells in industry: managing teams for profit'. *McGraw-Hill, Berkshire, England*, 1994.
82. K. Kurashige, Y. Yoshinari, S. Miyazaki, and Y. Kameyama. 'sequencing method for products in consideration of assembly time'. *Int. J. Prod. Eco.*, 60-61:565–573, 1999.
83. F. Kursawe. 'a variant of evolution strategies for vector optimisation'. *In Parallel Problem Solving from Nature. 1st Workshop, PPSN I, Lecture Notes in Computer Science, Berlin, Germany, Springer-Verlag*, 496:193–197, 1990.
84. A. Kusiak and W. S. Chow. 'decomposition of manufacturing systems'. *IEEE Journal of Robotics and Automation*, 4(5):457–471, 1988.
85. C. Y. Lee, M. Gen, and Y. Tsujimura. 'multicriteria assembly line balancing problem with parallel workstations using hybrid gas'. *In: Proceedings of the 3rd International Conference on Design and Automation (EDA99). Vancouver, Canada*, pages 115–122, 1999.
86. H. F. Lee and R. V. Johnson. 'a line-balancing strategy for designing flexible assembly systems'. *International Journal of Flexible Manufacturing Systems*, 3:91–120, 1991.
87. H. F. Lee and K. E. Stecke. 'an integrated design support method for flexible assembly systems'. *Journal of Manufacturing Systems*, 15(1):13–32, 1996.
88. Y. Y. Leu, L. A. Matheson, and L. P. Rees. 'assembly line balancing using genetic algorithms with heuristic-generated initial population and multiple evaluation criteria'. *Decision Sciences*, 25(4):581606, 1994.
89. M. Lucertini, D. Pacciarelli, and A. Pacifici. 'modeling an assembly line for configuration and flow management'. *Computer Integrated Manufacturing systems*, 11:15–24, 1998.
90. K. Mainzer. 'thinking in complexity: the complex of dynamics of matter, mind, and mankind'. *Springer-Verlag, Berlin Heidelberg*, 1994.
91. B. Malakooti and A. Kumar. 'a knowledge-based system for solving multi-objective assembly line-balancing problems'. *Int. J. Prod. Res.*, 34(9):2533–2552, 1996.
92. J.-F. Mare, B. Raucent, and A. Spineux. 'selection of assembly technique and equipment'. *Proceedings of ISATP'99, Porto, Portugal*, pages 393–398, 1999.

93. P. R. McMullen and G. V. Frazier. 'a heuristic for solving mixed-model line balancing problems with stochastic task durations and parallel stations'. *Int. J. Prod. Eco.*, 51:177–190, 1997.
94. P. R. McMullen and G. V. Frazier. 'using simulated annealing to solve a multiobjective assembly line balancing problem with parallel workstations'. *International Journal of Production Research*, 36(10):2717–2741, 1998.
95. G.J. Miltenburg and J. Wijngaard. 'the u-line balancing problem'. *Management Science*, 40:1378–1388, 1994.
96. J. Miltenburg. 'level schedules for mixed-model assembly lines in JIT production systems'. *Management Science*, 35(2):192–207, 1998.
97. J. Miltenburg and W. Zhang. 'a comparative evaluation of nine well-known algorithms for solving the cell formation problem in group technology'. *Journal of Operational Management*, 10(1):44–72, 1991.
98. V. Minzu and J.-M. Henrioud. 'approche systématique de structuration en postes des systèmes d'assemblage monoproduits'. *RAIRO-APII-JESA*, 31(1):57–78, 1997.
99. V. Minzu and J.-M. Henrioud. 'assignment stochastic algorithm in multi-product assembly lines'. *In: Proceedings of the 1997 International Symposium on Assembly and Task Planning (ISATP97), Marina del Rey, California*, pages 109–114, 1997.
100. Y. Monden. 'toyota production system: practical approach to production management'. *Industrial Engineering and Management Press, Atlanta*, 1983.
101. C. L. Moodie and H. H. Young. 'a heuristic method of assembly line balancing for assumptions of constant or variable work element times'. *J. Ind. Eng.*, 16(1):23–29, 1965.
102. T. Murata, H. Ishibuchi, and M. Gen. 'specification of local search directions in genetic local search algorithms for multi-objective optimisation problems'. *Proceedings of GECCO'99, Orlando USA*, pages 441–448, 1999.
103. J. L. Nevins and D. E. Withney. 'concurrent design of products and processes'. *McGraw-Hill, New York*, 1989.
104. P. D. T. O'Connor. 'practical reliability engineering'. *Third edition revised, John Wiley & Sons*, 1995.
105. A. Okano. 'computer-aided assembly process planning with resource assignment'. *Proceedings of the International IEEE Conf. on Robotics and Automation, Atlanta-Georgia*, pages 301–306, 1993.
106. G. Pahl and W. Beitz. 'engineering design: a systematic approach'. *Springer-Verlag, Berlin Heidelberg*, 1996.
107. V. Pareto. 'cours d'economie politique'. *F. Rouge, Lausanne*, I and II, 1988.
108. H. R. Parsaei and W. G. Sullivan. 'concurrent engineering: contemporary issues and modern design tools'. *Chapman and Hall, London, England*, 1993.
109. J. H. Patterson and J. J. Albracht. 'assembly line balancing: zero one programming with Fibonnaci search'. *Oper. Res.*, 23:166–172, 1975.
110. F. Pellichero. *'Computer-aided choice of assembly methods and selection of equipment in production line design'*. PhD thesis, Université Libre de Bruxelles, 1999.
111. Y. Peng. 'the algorithms for the assembly line balancing problem'. *Internal Report, CRIF Industrial Automation, Belgium*, 1991.
112. C. Peterson. 'a tabu search procedure for the simple assembly line balancing problem'. *Proceedings of the Decision Science Institute Conference, Washington, DC*, pages 1502–1504, 1993.

113. F. Petit. *'Interactive design of a product and its assembly system'.* PhD thesis, Université Catholique de Louvain, 1999.
114. A. Pinnoi and W. E. Wilhelm. 'assembly system design: a branch and cut approach'. *Management Science*, 44(1):103–118, 1998.
115. P. A. Pinto, D. G. Dannenbring, and B. M. Khumalawa. 'assembly line balancing with processing alternatives: an application'. *Management Science*, 29(7):817–830, 1983.
116. M. Pirlot. 'general local search in combinatorial optimization: a tutorial'. *Belgian Journal of Operations Research, Statistics and Computer Science*, 32(1):7–67, 1993.
117. S. G. Ponnambalam, P. Aravindan, and N. G. G. Mogileeswar. 'assembly line balancing using multi-objective genetic algorithm'. *In: Proceedings of CARS&FOF98. Coimbatore, India*, pages 222–230, 1998.
118. I. C. Praça and C. Ramos. 'multi-agent simulation for balancing of assembly lines'. *Proceedings of ISATP'99*, pages 459–464, 1999.
119. R. Rachamadugu and B. Talbot. 'improving the equality of workload assignments in assembly lines'. *Int. J. Prod. Res.*, 29(3):619–633, 1991.
120. H. K. Rampersad. 'integrated and simultaneous design for robotic assembly'. *John Wiley & Sons, London*, 1994.
121. B. Rekiek, P. De Lit, and A. Delchambre. 'designing mixed-model assembly lines'. *Special Issue of the IEEE Transactions on Robotics and Automation*, 16(3):268–280, 2000.
122. B. Rekiek, P. De Lit, and A. Delchambre. 'hybrid assembly line design and user's preferences'. *International Journal of Production Research*, 40(5):1095–1111, 2002.
123. B. Rekiek, P. De Lit, and A. Delchambre. 'evolutionary approach to design assembly lines'. *Submitted to Global Optimization Selected Case Studies edited volume*, 2006.
124. B. Rekiek, P. De Lit, A. Delchambre, F. Pellichero, E. Falkenauer, T. L'Eglise, and P. Fouda. 'concurrent engineering approach to design of assembly lines'. *Proceedings of CARS & FOF'2000, Port of Spain, Trinidad Tobago*, 1:333–340, 2000.
125. B. Rekiek, P. De Lit, F. Pellichero, E. Falkenauer, and A. Delchambre. 'applying the equal piles problem to balance assembly lines'. *Proceedings of the ISATP'99, Porto, Portugal*, pages 399–404, 1999.
126. B. Rekiek, P. De Lit, F. Pellichero, T. L'Eglise, P. Fouda, E. Falkenauer, and A. Delchambre. 'evolving to integrate logical and physical layout of assembly lines'. *Proceeding of the 4th International Conference on Engineering Design and Automation (EDA'2000), Orlando, USA*, 2000.
127. B. Rekiek, P. De Lit, F. Pellichero, T. L'Eglise, P. Fouda, E. Falkenauer, and A. Delchambre. 'a multiple objective grouping genetic algorithm for assembly lines design'. *Journal of Intelligent Manufacturing*, 12(6):467–485, 2001.
128. B. Rekiek and A. Delchambre. 'ordering variants and simulation in multi-product assembly lines'. *Proceedings of the VR-Mech'98, Brussels, Belgium*, pages 49–54, 1998.
129. B. Rekiek and A. Delchambre. 'assembly line balancing and resource planning: what is done and what is still missing'. *Proceedings of CARS & FOF'2001, Durban, South Africa*, pages 86–93, 2001.
130. B. Rekiek and A. Delchambre. 'hybrid assembly line design'. *Proceedings of ISATP'2001, Fukuoka, Japan*, pages 73–78, 2001.

131. B. Rekiek, A. Dolgui, A. Delchambre, and A. Bratcu. 'state of art of optimization methods for assembly line design'. *Annual Reviews in Control*, 26(2):163–174, 2002.
132. B. Rekiek, E. Falkenauer, and A. Delchambre. 'multi-product resource planning'. *Proceedings of the ISATP'97, Marina del Rey California, USA*, pages 115–121, 1997.
133. B. Rekiek, E. Falkenauer, and A. Delchambre. 'two problems in design and operation of multi-product assembly lines: line balancing and ordering variants'. *Proceedings of the CARS & FOF'98, Coimbatore, India*, pages 234–250, 1998.
134. B. Rekiek, F. Pellichero, P. De Lit, E. Falkenauer, and A. Delchambre. 'towards physical layout of assembly lines'. *Published as paper in the book entitled 'Progress in Simulation, Modeling, Analysis and Synthesis of Modern Electrical and Electronic Devices and Systems', World Scientific Engineering Society*, pages 307–312, 1999.
135. B. Rekiek, F. Pellichero, P. De Lit, T. L'Eglise, E. Falkenauer, and A. Delchambre. 'balancing and resource planning for assembly lines: the gap between theory and practice'. *Proceedings of the CPI'99, Tangier, Morocco*, pages 239–248, 1999.
136. B.J. Ritzel, J. W. Eheart, and S. Ranjithan. 'using genetic algorithms to solve a multiple objective groundwater pollution containment problem'. *Water Resources Research*, 30:1589–1603, 1994.
137. R. Romanowicz. *'A tool for an efficient comparison of scheduling methods'*. PhD thesis, Swiss Federal Institute of Technology of Lausanne (EPFL), 1997.
138. R. S. Rosenberg. *'Simulation of genetic populations with biochemical properties'*. PhD thesis, University of Michigan, Ann Harbor, Michigan, 1967.
139. B. Roy. 'classement et choix en présence de points de vue multiples (la méthode electre)'. *Revue Franaise d'Informatique et de Recherche Oprationnelle*, 8:57–75, 1968.
140. J. Rubinovitz and J. Bukchin. 'ralb a heuristic algorithm for design and balancing of robotic assembly lines'. *Annals of the CIRP*, 42:497–500, 1993.
141. I. Sabuncuoglu, E. Erel, and M. Tanyer. 'assembly line balancing using genetic algorithms'. *Journal of Intelligent Manufacturing*, 11:295–310, 2000.
142. E.D. Sacerdoti. 'a structure for plans and behavior'. *Stanford Research Institute, Elsevier, Amsterdam, Netherlands*, 1977.
143. M. E. Salveson. 'the assembly line balancing problem'. *J. Ind. Engng.*, 6(3):18–25, 1955.
144. B. R. Sarker and H. Pan. 'designing a mixed-model assembly line to minimise the costs of idle and utility times'. *Computers Ind. Engng.*, 34(3):609–628, 1998.
145. J. D. Schaffer. 'multiple objective optimisation with vector evaluated genetic algorithms'. *In Genetic Algorithms and their Applications: Proceedings of the First ICGA, Lawrence Erlbaum*, pages 93–100, 1985.
146. A. Scholl. 'balancing and sequencing of assembly lines'. *Heidelberg: Physica, Second edition*, 1999.
147. A. Scholl and R. Klein. 'salome: a bidirectional branch and bound procedure for assembly line balancing'. *INFORMS Journal on Computing*, 9:319–334, 1997.
148. H.-P. Schwefel. 'numerical optimisation of computer models'. *Great Britain: John Wiley & Sons*, 1981.

149. H.-P. Schwefel. 'evolution and optimum seeking'. *New York, NY: John Wiley*, 1995.
150. G. A. Süer. 'designing parallel assembly lines'. *Computers Ind. Eng.*, 35(3-4):467–470, 1998.
151. R. Sedgewick. 'algorithms'. *Addison-Wesley Publishing*, 1984.
152. P. Sen and J-B. Yang. 'multiple criteria decision support in engineering design'. *Springer-Verlag*, 1998.
153. K. Shimokawa, U. Jrgens, and T. Fujimoto. 'transforming automobile assembly: experience in automation and work organization'. *Springer-Verlag, Berlin Heidelberg, Germany*, 1997.
154. A. Shtub and El E. M. Dar. 'an assembly chart oriented assembly line balancing approach'. *Int. J. of Prod. Res.*, 28(6):1137–1151, 1990.
155. H. A. Simon. 'the sciences of the artificial'. *The MIT Press, Cambridge, MA, 3rd ed.*, 1981.
156. N. Srinivas and K. Deb. 'multiobjective optimisation using nondominated sorting in genetic algorithms'. *Evolutionary Computation*, 2:221–248, 1994.
157. L. Steinberg and K. Rasheed. 'optimizing by searching a tree of populations'. *Proceedings of GECCO'99, Orlando USA*, pages 1723–1730, 1999.
158. D. R. Sule. 'manufacturing facilities location, planning and design'. *PSW Publishing Company*, 1994.
159. G. Sureh and S. Sahu. 'stochastic assembly line balancing using simulated annealing'. *International Journal of Production Research*, 32(8):1801–1810, 1994.
160. G. Sureh, V. V. Vinod, and S. Sahu. 'a genetic algorithm for assembly line balancing'. *Production Planning and Control*, 7:38–46, 1996.
161. F. B. Talbot, J. H. Patterson, and W. V. Gehrlein. 'a comparative evaluation of heuristic line balancing techniques'. *Management Science*, 32:430–454, 1986.
162. T. Tamura, W. Wang, S. Fujita, and K. Ohno. 'line balancing for mixed-model assembly line with bypass sub-line'. *Proceedings of ICED'99, Munich, Germany*, 2:965–968, 1999.
163. T. B. To and U. Korn. 'mobes: a multiobjective evolution strategy for constrained optimisation problems'. *The Third International Conference on Genetic Algorithms Mendel'97, Brno Czech Republic*, pages 176–182, 1997.
164. J. A. Tompkins, J. A. White, Y. A. Bozer, E. H. Frazelle, J. M .A. Tanchoco, and J. Trevino. 'facilities planning'. *John Wiley & Sons Inc.*, 1996.
165. D. M. Tsai and M. J. Yao. 'a line-balanced-base capacity planning procedure for series-type robotic assembly line'. *Int. J. of Prod. Res.*, 31:1901–1920, 1993.
166. P. J. Van Laarhoven and E. H. Aarts. 'simulated annealing: theory and applications'. *D. Reidel, Dordrecht*, 1987.
167. D. A. Van Veldhuizen. *'Multiobjective evolutionary algorithms: classifications, analyses, and new innovations'*. PhD thesis, Department of Electrical and Computer Engineering. Graduate School of Engineering. Air Force Institute of Technology, Wright-Patterson AFB, Ohio, 1999.
168. M. B. Wall. *'A Genetic algorithm for resource-constrained scheduling'*. PhD thesis, Massachusetts Institute of Technology, 1996.
169. V. Wanet. *'Développement d'une bibliothéque de simulations d'eléments d'une ligne d'assemblage'*. PhD thesis, Travail de fin dtudes prsent en vue de lobtention du grade dingnieur civil mcanicien, Universit Libre de Bruxelles, Brussels, Belgium, 1999.

170. F. Z. Wang and R. C. Wilson. 'comparative analyses of fixed and removable item mixed-model assembly lines'. *IIE Transactions*, 18(3):313–317, 1986.
171. T. S. Wee and M. J. Magazine. 'assembly line balancing as generalized bin packing'. *Operations Research letters*, 1:56–58, 1982.
172. L. Wester and L. Kilbridge. 'the assembly line model-mix sequencing problem'. *Proceedings of the 3rd International Conference on Operations Research, Oslo*, pages 247–260, 1964.
173. P. B. Wilson and M. D. Macleod. 'low implementation cost IIR digital filter design using genetic algorithms'. *IEE/IEEE Workshop on Natural Algorithms in Signal Processing, Chelmsford, U.K.*, 4:1–8, 1993.
174. L. Zadeh. 'fuzzy sets'. *Information and Control*, 8:338–353, 1965.
175. E. Ziztler. *'Evolutionary algorithms for multiobjective optimization: methods and applications'*. PhD thesis, Swiss Federal Institute of Technology (ETH), Zurich Switzerland, 1999.

Index